Richard Norris

The Physiology and Pathology of the Blood

comprising the origins, mode of development, pathological and post-mortem

changes of its morphological elements in mammalian and oviparous vertebrates

Richard Norris

The Physiology and Pathology of the Blood
comprising the origins, mode of development, pathological and post-mortem changes of its morphological elements in mammalian and oviparous vertebrates

ISBN/EAN: 9783337390860

Printed in Europe, USA, Canada, Australia, Japan

Cover: Foto ©berggeist007 / pixelio.de

More available books at **www.hansebooks.com**

THE

PHYSIOLOGY AND PATHOLOGY
OF THE BLOOD:

THE ORIGIN, MODE OF DEVELOPMENT, PATHOLOGICAL AND POST-MORTEM CHANGES
OF ITS MORPHOLOGICAL ELEMENTS IN MAMMALIAN AND OVIPAROUS VERTEBRATES.

BY

RICHARD NORRIS, M.D., F.R.S.E.,

PROFESSOR OF PHYSIOLOGY, QUEEN'S COLLEGE, BIRMINGHAM;
VICE-PRESIDENT OF THE BIRMINGHAM PHILOSOPHICAL SOCIETY.

WITH MICRO-PHOTOGRAPHIC ILLUSTRATIONS.

"Truth is one: Error is manifold."

LONDON:
SMITH, ELDER, & CO., 15, WATERLOO PLACE.
1882.

PRINTED BY WRIGHT, DAIN, PEYTON, AND CO.,

AT THE HERALD PRESS, BIRMINGHAM.

To the Memory of

WILLIAM HEWSON,

ONE OF THE EARLIEST AND MOST ACUTE MICROSCOPICAL INVESTIGATORS OF

THE PHYSIOLOGY AND PATHOLOGY OF THE BLOOD.

THIS WORK

Is Gratefully Dedicated

BY

THE AUTHOR.

TABLE OF CONTENTS.

SECTION I.

ON THE EXISTENCE IN THE BLOOD OF A PREVIOUSLY UNKNOWN CORPUSCLE WHICH EXPLAINS THE ORIGIN OF THE RED DISC AND THE FORMATION OF FIBRIN.

PART III.—The granule sphere of Semmer (Körnerkugeln)—Found in horse's plasma at ice cold temperature—Coloured and larger than white corpuscles—Behaviour with reagents—Supposed

SECTION II.

FURTHER RESEARCHES ON THE THIRD CORPUSCULAR ELEMENT OF

MAMMALIAN BLOOD.

Advance in the physiology, pathology, and coagulation of the
blood impossible without a knowledge of this element—
Review of the position—Do these colourless discs exist in
normal unshed blood—All objectors now admit the existence
of these colourless discs in *shed* blood—Discussion hinges
altogether on the point as to whether they are *decolourised red
discs*—All the facts, without exception, are against this view—
1. Their presence can be demonstrated without reagents of any
kind, that is by simple mechanical arrangements—2. They

SECTION III.

ON THE MOST SUCCESSFUL METHOD OF STAINING THE CORPUSCLES OF THE FUGITIVE GROUP.

Difficulties connected with staining—No sharp line of demarcation owing to the corpuscles being a graduated series—Power to stain diminishes with assumption of hæmoglobin—Invisible or colourless discs stain blue, those containing a little hæmo-

SECTION IV.

AN EXAMINATION OF THE RESEARCHES OF M. HAYEM ON THE DEVELOP-
MENT OF MAMMALIAN BLOOD.

Title and date of Hayem's papers—Statement of his views and
examination of his methods—Small elements exist which are
neither red nor white corpuscles—These are the *germs* of the
red corpuscles and may be called *hæmatoblasts*—Avoids discus-
sing the origin of these elements—Seeks to ascertain the exact
anatomical constitution of the blood at birth—The so-called
hæmatoblasts are *visible* elements of the blood, and have
colour—Are seen best in fresh blood—Method of observing
these elements—High powers required—Very small bodies like
delicate, pale red corpuscles—Grow paler by loss of hæmoglobin—Change their form—Adhere to the slide and to the red
discs—Unite with each other and form wreaths or groups—
May be preserved for examination by cold (0° C.)—So
preserved, show themselves as thin, discoid, biconcave, slightly
coloured bodies—Osmic acid and bichloride of mercury
preserve them—Full description—Bear no resemblance to
white corpuscles—Present two parts or constituents—One
portion exudes, is viscous, and is the cause of their adhesiveness—Identification of these elements—Summary of the
various morphological elements contained in *shed* blood—Easy
to identify among these the elements mistaken by Hayem for
the germs of the red corpuscles—Modified and altered forms
of the younger blood discs—Comparison of the general properties of these elements with those of the discs of the
fugitive group—The discs of this group must be followed through
their transformations—Photograph 61 shows in one group the
entire series of blood discs—Characters of the youngest discs—
Power of retaining *form* in the ratio of colour—The younger

SECTION V.

ON THE MORPHOLOGICAL PRODUCTS OF THE BLOOD GLANDS.

SECTION VI.

ON THE ORIGIN, DEVELOPMENT, AND DESTINY OF THE WHITE BLOOD
CORPUSCLE.

SECTION VII.

ON THE IDENTIFICATION OF THE ADVANCED LYMPH DISC WITH THE
COLOURLESS DISC OF THE BLOOD.

SECTION VIII.

ON THE RÔLE OF THE RED BONE-MARROW IN THE FORMATION OF BLOOD.

ON THE RELATION OF THE MARROW OF THE ADULT MAMMAL TO ITS

BLOOD.

SECTION IX.

NATURE OF LEUKHÆMIA.

Discovered by Bennett and Virchow—Virchow first associated the
condition with enlargement of spleen and glands—Red discs
diminish proportionately with increase of white corpuscles—
This fact has never been explained—Condition of spleen and

SECTION X.

ON THE NATURE OF ANÆMIA.

SECTION XI.

ON THE RELATION WHICH THE PRODUCTS OF THE BONE-MARROW AND OF
THE SPLEEN OF THE ADULT OVIPARA BEAR TO ITS BLOOD.

SECTION XII.

ON THE ANALOGY AND RELATION WHICH SUBSIST BETWEEN THE DEVELOPMENT OF MAMMALIAN AND OVIPAROUS BLOOD.

APPENDIX.

LIST OF MICRO-PHOTOGRAPHS AND ILLUSTRATIONS.

PREFACE.

When in August. 1877, the discovery was made that there existed in the blood of mammals a large number of corpuscles, which had the same *colour* and the same *refractive index* as the liquor sanguinis, and which were therefore necessarily invisible in this liquid, all other laboratory work was suspended, and I and my assistant set ourselves earnestly to work with a view, either to verify or disprove what appeared to be so important a fact. Every day for five months was exclusively devoted to this purpose, and by the end of this time, we had not only in numerous new ways verified the original observation, but had also ascertained that these *new corpuscles* bore important relations on the one hand to the well-known red discs, and on the other to the corpuscles of the lymph and spleen. In addition it became apparent that the abnormal deviations of this newly-discovered element gave rise to *fibrin* in its various forms of films, networks, thrombi, and emboli, that they were, in fact, the cause of coagulation generally.

In February, 1878, having acquired a confidence, begotten of incessant self-distrust, and of constant reappeals to Nature, I made my first overtures towards the publication of the results, but from circumstances over which I had no control this was delayed till November 14th of the same year, on which date a paper was read to the Birmingham Philosophical Society, under the following title :—" *On the Existence in Mammalian Blood of a New Morphological Element which explains the Origin of the Red Disc and the Formation of Fibrin.*"

For several years I had been experimenting with the view of rendering micro-photography by the high powers available for the illustration of physiological and pathological investigations. I desired to do this for the reasons mentioned in the introductory remarks to Part I. of this work.

At the period of the discovery of the *new corpuscle* I had so far succeeded as to be able to produce with great rapidity, with the high powers of the microscope, negatives of sufficient intensity for printing delicate transparencies upon glass.* I determined, therefore, to photograph the results I had obtained with the blood, feeling sure that this would be the best means of promoting and making known my views, for I can in this way place before my readers exact representations of what I have myself seen under the microscope, and thus, though they may be unskilled, or but partially skilled in the use of this instrument, they will be able to understand the precise bearings of the question as easily and with the same security as if they had devoted considerable time to its use, for while the absolute and relative sizes and forms of objects are rigidly maintained, the varying yellow tints are transmuted into corresponding degrees of light and shade, which every eye can estimate with equal correctness.

Many delicate things can also be observed in the photographs, which, owing to sameness of tint and the glare of light in the microscope, can only be seen with difficulty, and in some cases not at all in the original specimens.

This is the first time that micro-photography with the high powers has been pressed systematically into the service of research, and in addition to affording a new method or reagent of great delicacy in relation to colour, and light and shade, its advantages over the ordinary methods of delineation are so

---

* These glass transparencies were first exhibited publicly at a soirée of the Royal College of Physicians, London, in July, 1879; and in August of the same year I gave two demonstrations of the subject at the meeting of the British Medical Association held in Cork. 1. On the new morphological element of mammalian blood and its relation to the development of the red blood disc. 2. On the rôle of this new element in fibrin formation and coagulation generally. I had early in 1878 succeeded in obtaining negatives of sufficient intensity to print satisfactorily upon paper. An abstract of these discoveries was published in the ninth edition of Carpenter's Physiology in the article on the blood, and later on, in 1880, a second paper was read to the Birmingham Philosophical Society, bearing the title, "Further researches on the third corpuscular element of mammalian blood."

palpable that its use seems likely to constitute a new epoch in
illustration, for there can be little doubt that it will in the long
run be adopted by all microscopic investigators who desire their
labours to have proper weight and permanent influence.

Photography appears to be the natural handmaid of
Microscopy, and there can be no doubt that its universal
adoption would lead to a rapid development of the sciences
which depend upon this instrument for their elucidation, as it
would enable each investigator to examine more perfectly, and
weigh more critically, the results submitted to his judgment by
fellow workers. It is well known that some of our leading
scientific societies entertain a great distaste for papers on microscopic subjects. This feeling has no doubt originated in the
great difficulties which beset such researches, for, in order to
verify and adjudicate upon them *honestly* and *properly*, it is
necessary to follow in the precise tracks of the investigator, a
thing almost impossible without the special education in the new
methods which the discoverer himself has alone achieved.
The difficulties are in reality of such magnitude as to render it
all but impossible, as things stand, for a wise and just decision
upon the value of such papers to be made. The general use of
micro-photography would undoubtedly in great part remove
this difficulty, and as there are many important regions of science
which can only be investigated by means of the microscope,
it is most desirable that this lamentable state of things
should soon cease to exist.

One of the great advantages, also, of this method consists
in the fact that comparisons of specimens can be made with the
greatest facility whenever any new idea or fresh conception
occurs to the mind. Results which at one time are meaningless,
become pregnant with interest when compared with some other
result, or when viewed from a different standpoint. We know
also that imperfect generalisations too frequently arise from
the basis of facts being meagre and insufficient. During
the past few years I have accumulated several thousand
photographs, which are imperishable records of the varia-

tions and changes of which the formed elements of the blood are susceptible ; and these, when repeatedly examined and compared with each other, continue to throw immense light upon some of the obscure regions of physiology and pathology. I am, therefore, hopeful that great advances will be made in our knowledge when these methods are better under-stood and commonly adopted.

As the views of the blood, set forth in these essays, rest upon the discovery of *a new anatomical fact*, they are necessarily novel ; and as, in addition, some of the experiments, if not difficult, require painstaking care and patient applica-tion, I do not anticipate for them a ready recognition or acceptance, but in the meantime console myself by the reflection that I have trodden the ground conscientiously, and have laid down to the best of my ability the landmarks by which others may become familiar with the road ; and I am gratified that I can now feel myself at liberty to press steadily forward towards the important practical issues which I already see opening up in my path. I desire to add, that as my studies in this direction are likely to extend over some years, I shall feel obliged for any candid and generous criticisms with which persons working in the same field of labour may favour me, and for any help tending in any way to make the settlement of these vexed questions one of the many conquests of the century in which we live.

Many requests have been made to me by medical friends and others engaged in teaching for an opportunity of witnessing the various new methods of experimenting on the blood. Nothing would give me more pleasure than to accede to these wishes, but my time is so much occupied that I find it impossible to respond to individual demands. To meet the difficulty a room has been set apart at the Institute of Scientific Research, Broad Street, Birmingham, in which the apparatus, specimens, and photographs are so arranged and classified as to enable persons to quietly study the question in all its bearings at their own leisure.

In conclusion, I have to thank many scientific friends, not only for sound advice, but also for the moral support and encouragement which they have accorded me under somewhat trying circumstances, and more particularly is my gratitude due to my friend Mr. Henry Fulford, to whom I am deeply indebted for much literary assistance, and also for generous and substantial aid in carrying out the work, which, on account of the novelty of the method of illustration, has been attended with a cost which would have been very burdensome to me as a private individual.

At the moment of going to press my attention has been drawn to a paper by Professor Bizzozero, of Turin, just published (January 14th, 1882) in the " Centralblatt für die medicinischen Wissenschaften " as a " preliminary notice " in the shape of an " original communication," bearing the title " Ueber einen neuen Formbestandteil des Säugetierblutes, und die Bedeutung desselben für die Thrombosis und Blutgerinnung überhaupt."*

As the statements made in this paper are *essentially* and almost *verbally* the same as those which appear in my papers of 1878 and 1880 to the Birmingham Philosophical Society, I shall, to facilitate comparison, take the liberty of placing them in contrast side by side.

BIZZOZERO, 1882.	NORRIS, 1878 and 1880.
TITLE OF PAPER.	TITLES OF PAPERS.
" *On a new element of Mammalian Blood and the part it plays in the production of thrombi and coagulation generally.*"—"Centralblatt," January 14, 1882.	" *On the existence in Mammalian Blood of a new Morphological Element which explains the origin of the Red Disc, and the formation of Fibrin.*"—Nov.14, 1878—Proc. Birm. Phil. Soc.
	" *Further Researches on the Third Corpuscular Element of Mammalian Blood.*" — June 10, 1880—"Proc.Birm.Phil.Soc."
" If the course of the circulation is watched in the mesentery of	" From the above observations the author considers that two

---

* *Vide* Original Paper in Appendix.

chloralised rabbits and guinea-pigs, there are seen, besides the ordinary red and white corpuscles, third elements, very pale, oval or round, disc-shaped, or lenticular bodies, one-half or one-third the diameter of the red corpuscles among which they are scattered."

conclusions are justifiable. 1st, That there exists in the blood of mammalia, in addition to the well-known red and white corpuscles (*Vide* Plate II., Photographs 1 and 2), colourless, transparent biconcave discs of the same size as the red ones. 2nd, Between these two kinds of biconcave discs, others having every *intermediate* gradation of colour are demonstrable."

" The *origin* of these new corpuscles is treated of in the second part of this paper."—*Proc. Birm. Phil. Soc., 1878, p. 16.*

" As in some of the examples yielded by the previous methods, there appeared to be indications that the *normal* form of the transparent colourless corpuscles was that of the unaltered red corpuscle, viz., a biconcave disc, it was decided to attempt the preservation of their true form by separating the glasses over a shallow pan filled with the vapour of osmic acid, it being well known, as pointed out by Schultze, that this vapour possesses the property of preventing change in the form of the red corpuscle. The series of Photographs 14, 15, and 16, on Plate IV., were obtained by this method, and are remarkable groups for displaying the fact, that corpuscles exist of every gradation of tint, from those which are perfectly colourless to the fully cruorised red disc. In Photograph 16 we have also evidence that the corpuscles which are freest from colour possess in their original state a biconcave form the most decisive results, however, as to the biconcave form of the new corpuscle have been obtained by the use of cold and osmic acid vapour in conjunction."—*Proc. Birm. Phil. Soc., 1878, p. 10.*

" These bodies have hitherto escaped notice, probably because they are so colourless and translucent, less numerous than the red, and less visible than the white corpuscles."

" If we place upon the tip of the finger a minute drop of saturated solution of salt, and prick through it, so that the blood may flow directly into the saline solution, the refractive power of the liquor sanguinis is modified, and it is found that if we run this mixture of salt and blood between glasses prepared according to the packing method before described, we can then see the *outlines* of the transparent corpuscles and the clear spaces which have hitherto been supposed to consist of liquor sanguinis only, are observed to teem with these corpuscles as in Plate II., Photograph 6."—*Proc. Birm. Phil. Soc., p. 6.*

" Although the blood may be said to swarm with these transparent corpuscles they escape observation, because their *colour* and *refractive index* coincide exactly with those of the liquor sanguinis."—*Proc. Birm. Phil. Soc., p. 6.*

" These corpuscles are to be observed also in freshly-drawn blood, for the most part aggregated around the colourless corpuscles, or ascending to the upper layer, they adhere to the cover-glass."

" *Specific Gravity.*—Like the white corpuscle, they are lighter than the red, and have a tendency constantly to rise to the surface of the blood, consequently the largest numbers are always seen to attach themselves to the upper glass in preference to the lower, and especially if time is allowed them to rise. This, no doubt, has something to do with the *buffy coat.*"—*Proc. Birm. Phil. Soc., 1880, p. 266.*

" These corpuscles are extremely fugitive in their character when the blood is shed, and are rendered more so by dilution of the plasma."—*Proc. Birm. Phil. Soc., 1878, p. 13.*

" They change, however, with great rapidity, rapidly becoming granular, and appear to be the source of the small granule masses which have been described by many observers."

" The first corpuscles to attach themselves to foreign bodies, such as glass, are the uncoloured, and very slightly coloured, smooth corpuscles, the invisible corpuscles of the blood. But these are easily overlooked, for they are so delicate and fragile that when they do not melt down, so to speak, they almost invariably break up into molecules or spheroidal granules, however gently the surplus blood may be withdrawn by capillarity from off and around them; that is to say, however carefully the glasses may be separated from each other. Another mode in which they frequently elude observation is by fusing and laying themselves down in a delicate film upon the slide and cover-glass, and they often seem in this way to furnish a basis of adhesion for the more cruorised corpuscles, for mosaic groups of these are constantly found to have a delicate layer of the non-cruorised corpuscles beneath and around them."—*Proc. Birm. Phil. Soc., 1878, p. 9.*

" *Granulation.*—The red corpuscles rarely undergo granulation, but these can scarcely be prevented doing so. In this respect they are like the ordinary white corpuscle, to which body they in fact assimilate in *all* their properties. These granules sometimes result from the breaking up of single corpuscles, and at others from the breaking up of groups or fused masses of them." —*Proc. Birm. Phil. Soc., 1880, p. 216.*

" The corpuscles can be preserved unaltered in form by more prolonged examination by certain reagents, as for instance by a solution of sodium chloride, tinted with methyl-violet."

"The plan I have found successful has been to use saturated solutions of these salts, more especially sodium chloride, as the basis of the staining fluid; that is to say, I have used pigments which could be retained in solution, in water saturated with these salts."—*Proc. Birm. Phil. Soc., 1878, p. 13.*

" *Relation to Stains.*—I have already stated in my former paper that these corpuscles, when in the liquor sanguinis, stain with carmine, and the red ones with aniline. I have found more recently that they may be readily stained by a weak solution of aniline blue, in three-quarter per cent. solution of common salt."—*Proc. Birm. Phil. Soc., 1880, p. 217.*

" They are to be found also in human blood, but they undergo

"In obtaining the specimens of blood for examination the

alterations with extreme rapidity, and the best method of observing them has been found to be to place a drop of the above solution over the puncture, and then squeezing the blood out and immediately examining it under the microscope."

following method should be adopted in all cases :—Place upon the end of the finger a small drop of the staining fluid, and with a needle prick the finger through this drop, so that the blood may, when the finger is squeezed, flow directly into the liquid which has the double property of both *preserving and staining*. After well mixing with the needle on the end of the finger, the blood may be allowed to flow by capillarity between the cover-glass and slide for examination."—*Proc. Birm. Phil. Soc., 1878, p. 14.*

"A drop of this stain being placed upon the end of the finger, the latter is pricked through the drop, so that the blood may come into immediate contact with the saline solution the moment it is shed, and the corpuscles be thus prevented from undergoing change."—*Proc. Birm. Phil. Soc., 1878, p. 15.*

———

"The new elements constitute the chief part of the white clots in mammalia, since they give rise to the granular material which is seen between the white corpuscles, and which has hitherto been ascribed to the degeneration of fibrin." ·

"The time at which coagulation occurs in a given drop of blood corresponds closely to that at which these new elements present degenerative changes."

———

"Finally, the general behaviour of these corpuscles after the blood is shed, their tendency to break up into granules, to lay themselves down as delicate films, to form networks, in a word, their fibrin-forming property, is totally opposed to the conception that they were once red discs."—*Proc. Birm. Phil. Soc., 1878, p. 18.*

"The discovery of a third corpuscle has thrown great light on the question of the coagulation of the blood, and of fibrin formation generally. To avoid complication this subject will receive separate treatment. It may, however, be briefly stated here that on the basis of their behaviour when the blood is shed, the biconcave discs are divisible into two groups—a fugitive and a permanent group—and that the changes which take place in the former determine coagulation."—*Proc. Birm. Phil. Soc., 1878, p. 13.*

"In many cases the invisible corpuscles are still sufficiently distinct to show that these "fibrin or plasmine pools " originate from them. There is reason to think that every part of the background of this specimen is covered with liquid of corpuscular origin."

"Plate V., Photograph 22,shows invisible and subcruorised corpuscles in the act of spreading, fusion, and disintegration. It represents a still more advanced stage in plasmine formation."

" The disintegrative changes which take place in these corpuscles give rise to the formation of fibrin in the blood, and the fibrin which is formed in the lymph has its origin in similar changes in the gland corpuscles." —*Proc. Birm. Phil. Soc., 1878, p. 11.*

"Photograph 60, Plate XI.,gives an example of the direct conversion of the *colourless discs* into *fibrin* without passing through the stage of granulation. This is a modified mode of action of the process of annulation."—*Proc. Birm. Phil. Soc., 1880, p. 214.*

" These primary groups are often to be seen undergoing conversion into fibrin. The corpuscles of these groups are *de facto* fibrin, and the delicate fibres and layers which appear on glass-slides are due, first, to the extension of these granulations into fibres, or to annulation of the entire corpuscle ; or, secondly, to the spreading and laying down of these corpuscles into films. When blood is completely defibrinated these corpuscles and their granules entirely disappear, and can no longer be shown by any of my methods."—*Proc. Birm. Phil. Soc., 1880, p. 214.*

" These three classes of corpuscles are all capable of undergoing similar changes, but with different degrees of facility, and on this account these changes are commonly seen in the primary or fugitive group only."

" These changes are of the nature of *fusion*, of *granulation*, and of *fibrillation*, and groups of each class may be shown in which these changes have occurred or

are taking place."—*Proc. Birm.
Phil. Soc., 1880,* p. *213.*

" The application of a delicate
photo-chemical test, such as is
afforded us by photography, indubitably shows the existence of
a regularly graduated series,
from a colourless to a deep
yellow disc. Of these the colourless and the more faintly-tinted
ones range themselves together
on the unstable or fibrin side,
and the more strongly-tinted on
the stable or permanently corpuscular side ; in other words,
the stability of the blood corpuscle is directly proportionate
to its degree of cruorisation, and
the flickering or diffused edged
corpuscles mark the point at
which the biconcave discs become
converted into fibrin when the
blood is shed."—*Proc. Birm. Phil.
Soc., 1880, p. 215.*

" It will be remembered that
A. Schmidt asserted that the
coagulation of the blood is effected
by the white corpuscles, which,
by their destruction, yield the
granules, and so constitute a
considerable part of the substance of the clot. The formation of the clot is due not to the
white corpuscles, but to the new
elements."

THE GRANULE SPHERE
(KÖRNERKUGELN).

" In *Pfluger's Archives* for
November, 1875, will be found a
paper by Alexander Schmidt, in
which publicity is given to certain
observations made on the blood
by Semmer. This investigator
examined microscopically the
plasma of horse's blood which
had been prevented from coagulating by means of cold, and
arrived at the following conclusions :—1st, That there exists in
the plasma of horse's blood,
which has been allowed to subside at an ice-cold temperature,
yellow and red granule balls or
corpuscles, (rothe körnerkugeln,)
which are considerably larger
than the ordinary colourless corpuscle of the blood. 2nd, When
kept for a few hours 'these
corpuscles disintegrate into white
granular heaps ' (farblose körnerhaufen)."

" It is suggested by Semmer
that these granule balls or
spheres occupy an *intermediate
position* between the ordinary
white corpuscle and the red disc ;
that they are in fact the transi-

tion stages in the development of the red disc.

It is assumed, therefore, that they exist in *perfectly fresh* blood, *i.e.*, in blood circulating in the vessels, but disappear immediately when the blood is shed, going in some way to form fibrin."

"When blood is allowed to coagulate, the granule balls of Semmer mainly disappear, being simply the more coloured corpuscles of the *fugitive group*. They melt down into fibrin without undergoing granulation, but when by the use of cold this is prevented, they undergo granulation, and show themselves as coloured granule spheres. When the corpuscles first break up these granules are coloured, but they subsequently, *i.e.*, in 'a few hours,' give up their colour to the liquor sanguinis and appear white."—*Proc. Birm. Phil. Soc., 1878, p. 26.*

"The fluids which retard or prevent coagulation—solution of carbonate of soda, or sulphate of magnesia, for instance—also hinder the granular transformation of the new corpuscles."

"It is well known that the coagulation of the blood can be entirely prevented by means of saturated solutions of neutral salts. I have ascertained that this is due to the power of these substances to maintain the integrity of the invisible and subcruorised corpuscle, which is the true fibrin stroma."—*Proc. Birm. Phil. Soc., 1878, p. 13.*

"For after I had acquired the knowledge that these corpuscles were the *fibrin factors*, it occurred to me that, as neutral saline solutions prevented coagulation, they might do so by hindering physical changes in these corpuscles, and on examination I found this view to be correct."—*Proc. Birm. Phil. Soc., 1880, p. 220.*

What relation then do the corpuscles seen by Bizzozero in the circulating blood bear to those which I have collectively described as *the fugitive group of discs ?* I have shown in my early papers that the so-called lymph corpuscles are in reality discs. They are *smaller*, but somewhat *thicker*, than the blood discs, and are of two kinds. I have designated them the primary and the advanced lymph discs. The latter are the free nuclei of the former, and, *in the main*, enter the blood in such a forward state of development as to become *at once* its smooth invisible, colourless discs; but a few less perfectly elaborated present themselves in the blood as very delicate, translucent, visible discs, free from hæmoglobin. These, however, gradually pass into the invisible state. On the other hand, a few of the primary discs before referred to, which have escaped decapsulation in the blood-glands, come over into the blood. These are more visible than the foregoing, and have a slightly larger size. They do not pass into the invisible state, but, on the contrary, undergo development into the uninuclear white corpuscle, of which bodies there exist every variety of size between this primary cell and the multinuclear white corpuscle. In histological constitution the *fugitive group of discs* is entirely nuclear, beginning with the advanced lymph disc (a small proportion of which, as before said, exist in the blood as very delicate, visible bodies, but in the main as the invisible discs), and ending in the green, lustrous or diffused-edged corpuscles. The accidentally visible element of this group, which antecedes the invisible disc, is the corpuscle seen in the circulating blood by Bizzozero. Its existence is but the indication of the presence of a far more numerous series of corpuscles which are wholly invisible to the eye, but some of which are nevertheless capable of being photographed. In this *fugitive group of discs* all the elements exist which are essential to the formation of fibrin and the coagulation of the blood. These elements, studied from their chemical aspect, have received the well-known designations of fibrinogen, fibrinoplastin, and ferment, and may exist together or separately from each other. A single, somewhat developed, fugitive young blood disc, being convertible into fibrin, must contain them all, but such a disc in its physical and chemical constitution is but a type of the whole series of the *fugitive discs*, for in its interior it contains material of the nature of the primary invisible discs, and on its exterior of the coloured discs (*vide* Section IV.) Liquids such as serum and hydroclee fluid are *selective* by the mode of their origin—the former containing corpuscular derivatives which correspond to fibrinoplastin, and the latter such as correspond to fibrinogen. The fibrin which forms in the *lymph* is a product of the *advanced lymph discs*, while the ordinary clot which forms in *blood* in a state of quiescence is formed entirely of these and the invisible corpuscles; but some of the latter, although they do not contain sufficient colour to enable them to be seen, respond nevertheless to the more delicate colour-test afforded by photography. The formation of fibrin in the blood of the ovipara proceeds upon precisely the same principles from the degenerative changes in the younger corpuscles, especially of those with the *clear invisible margins*, which latter break away from their nuclei, and becoming diffused in the liquid pass readily through filtering paper, etc. After spontaneous coagulation thus brought about, it is well known that an amount of fibrin greater than that yielded *spontaneously* can be obtained by treating the corpuscles which have been allowed to subside in defibrinated blood or those on the filter with distilled water, which sets free the fibrin factors. In this case, therefore, we also have evidence that the corpuscles are the source of the fibrinogen, of the fibrinoplastin, and the ferment. The elliptical corpuscles of the ovipara are with those of the mammal equally capable of division into a fugitive and permanent group, based upon their behaviour when the blood is shed. For further information I must refer my readers to the Appendix, in which this matter of priority is more fully discussed, and leave them to judge whether I have adduced sufficient evidence to show that the statements made by Professor Bizzozero in 1882 had been already anticipated in my published writings of 1878 and 1880.*

Institute of Scientific Research,
   Birmingham, 1882.

---

* *Vide* Abstract from the Minutes of the Birmingham Philosophical Society, February 9th, 1882, in Appendix.

INTRODUCTION.

Probably no part of microscopical anatomy has exercised the powers or exhausted the patience of physiologists more than the question of the origin and mode of development of the morphological elements of the blood. These bodies are universally known under the designation of the red and white corpuscles. The former were first observed by Swammerdam in 1658,[*] and the latter by Hewson in 1773,[†] or 115 years later. The red corpuscle has, therefore, been known for 223 years, and the white one for 108 years ; but, notwithstanding the length of time these bodies have been the subject of investigation, no settled convictions have been arrived at, either as to their origin, mode of development, or the relation, if any, which they may hold to each other. Since

---

[*] These minute bodies were first seen in the year 1658 by Swammerdam. His observations, however, were not published till a century later. In 1661 Malpighi published his discovery of the blood corpuscles of the hedgehog. He erroneously regarded them as globules of fat. Leeuwenhoek, in 1673, detected them in human blood, and from this time the study of these bodies commenced in earnest. Hewson, in 1770, showed that the human red corpuscles in their normal state were not globules, but "in reality flat bodies." Dr. Young, in 1818, inferred that a depression existed on their flat surfaces—that, in fact, they were biconcave discs, this was finally determined by Dr. Hodgkin and Mr. Lister in 1827.

By the combined labours of Dr. Hodgkin, Mr. James Jackson Lister, Professor Gulliver, and Mr. Wharton Jones, it has been definitely determined that the mammalian red blood corpuscle does not possess a nucleus. As the result, then, of researches extending over two hundred years, we are in a position to affirm with certainty that the mammalian red blood corpuscles in their normal state are non-nucleated biconcave discs of a reddish yellow colour, having a diameter of about 1-3200th, and a thickness of about 1-12,000th of an inch.

[†] Secondly, we have proved that vast numbers of (central?) particles made by the thymus and lymphatic glands are poured into the blood vessels through the thoracic duct, and if we examine the blood attentively we see them floating in it. Hewson's Works, Sec. 98, p. 282.

the year 1846* no important effort has been made till very recently to grapple again with the subject. The most important

* The researches of Wharton Jones are published in the Philosophical Transactions for 1846. This author held that the red corpuscle was developed in the blood current from the white corpuscle, which he regarded as having three phases, that of granule-cell, of nucleated cell, and of free cellæform nucleus. " In the invertebrata and oviparous vertebrata the blood corpuscle exists only in the first two phases. The red corpuscle of the oviparous vertebrata belongs to the phase of nucleated cell; is, in fact, a white corpuscle which has attained colour. It is found in all the three phases in mammalia. The phase of free cellæform nucleus is the *colourless stage of the* well-known mammalian red corpuscle." This view has been before the world for more than a quarter of a century, and has enjoyed the advantage of illustrious support, but has failed entirely to win the adhesion of physiologists either in this country or abroad, because it has been impossible to trace the transformation of the liberated *naked nuclei* into *coloured discs*, or to give sound reasons why this could not be done.

This view, notwithstanding, is a close approximation to the truth, so far as the process is concerned, which in this work has been designated the *minor mode* of blood-making. The theory, indeed, is correct, but its author erred in considering that he could see the steps of the process, and trace in the blood the *naked nuclei* of the white corpuscles. Like the nuclei of the gland and splenic corpuscles, these bodies pass through the stage of *invisible colourless discs*, and do not reappear until they have gained a tint in advance of that of the liquor sanguinis. Wharton Jones appears to have mistaken the smallest kind of white blood corpuscles for the liberated nuclei of the larger ones. These are, however, not *naked nuclei*, for by the " method of osmosis " these bodies can be shown to possess a cell wall, and to be, in fact, *primary* lymph corpuscles. Be this as it may, it is not possible to trace *visually* the liberated nuclei of the white corpuscles through their *entire course*, because at one stage they have the refractive index and colour of the liquor sanguinis.

The view advanced by Kölliker in 1852, was also approximatively correct, but in this case the process, which in this work has been described as the *major mode* of blood-making, was the one concerned. It is obvious that he had no confidence in the previous doctrine which regarded the red disc as the *coloured* nucleus of the white blood corpuscles, for he says : " Though the microscopical investigation of the blood shows that it invariably contains a certain number of *larger pale cells*, with several nuclei or a single nucleus, disintegrated by acetic acid, of which, although they are certainly derived from the chyle or are *metamorphosed elements* of it, it is perhaps *impossible* to suppose that they ever become blood cells."— *Human Histology*, Vol. ii., p. 345.

Subsequently he gives his own view in the following words : " Notwithstanding the great pains specially devoted to the discovery of the origination of the blood globules after birth and in the adult, it still papers which the interval has produced are those of Bizzozero[*] and Neumann,[†] who, it must be admitted, have succeeded in showing that the red bone-marrow is a source of *lymphoid cells.*

These researches have, however, thrown no satisfactory light upon the genesis of the red corpuscle,[‡] and, as a conse-

---

remains one of the most obscure parts in the history of the blood cells. In my opinion, however, the notion which assumes that the red blood cells proceed from the *smaller chyle corpuscles* which lose their nuclei, become flattened, and have hæmatin produced in them, is the one most deserving of credit.   These cells are about the same size as the blood globules, or even rather smaller, have the same kind of membrane as the latter, are flattened, and not unfrequently of a yellow colour, and consequently may, as we see in the colourless blood cells of the embryo, pass without any considerable change into coloured cells.   Where and how this takes place no one has seen, and notwithstanding all the trouble and care that I have devoted to the subject, I have never noticed a nucleated coloured blood cell in the adult mammal.

"I maintain," he says, "their origination from the lymph corpuscles, and in order to explain the reason why the transition itself has not yet been observed, I broach the supposition that it may take place too rapidly to be in any way obvious with our means of observation."—*Human Histology*, Vol. II., Page 344.

Between the small lymph corpuscle in the thoracic duct and the red blood corpuscle, Kölliker saw nothing.   *There was a gap which could not be bridged over.*   He therefore inferred that the *small* lymph corpuscle became *suddenly* converted into the red corpuscle.

Kölliker needed two facts to enable him to place his view of the origin of the red corpuscle upon a solid foundation : 1. The knowledge that the small lymph corpuscles were simply *smooth nuclei* by the time they reached the thoracic duct ; and, 2, the fact of the existence in the blood of a *colourless disc* which *gradually* became coloured.

We see, therefore, that the views of both these distinguished histologists were in the main correct, but that they referred to *two distinct modes* of blood production, and that the knowledge of the invisible colourless stage of the blood disc was essential to the completion of either theory.

* 1865, 1868.—Sul midollo della ossa Napoli, 1869.

† 1868.—Du rôle de la moella des os dans la formation du sang. (Extrait.)   Page 1112, Comptes rendus, 1869.

‡ Both physiologists and pathologists have of late years become painfully conscious that we are destitute of all accurate knowledge respecting the matter.   Thus Virchow says : "The whole history of the red blood corpuscles is still invested with a mysterious obscurity, inasmuch as no positive information has, even at the present time been obtained with regard to the origin of these elements.   We know that in the first months of the existence even of the human embryo, divisions

quence, still more strenuous efforts have been made during the past few years to tear down the veil. Striking boldly, but somewhat rashly, clear of the old lines, Hayem, in France,[*] has endeavoured to show that the red corpuscles are entirely independent of the white corpuscles and of the lymph cells, and that they arise within the blood itself from minute bodies which he has described and designated *hæmatoblasts.* On the other hand, Rindfleisch, in Germany,[†] smarting under the humiliating sense of our profound ignorance, which he describes as a sore in our scientific manhood, has sought to support and develop the views of the medullary school, by affirming that cells similar to red nucleated embryonic cells are formed in the red bone marrow, the nuclei of which do not, as previously

take place in the cells, whereby an increase in the number of those present in the blood itself is produced. But after this time all is obscure, and this obscurity indeed corresponds pretty exactly with the period at which the corpuscles in the blood of man and the mammalia cease to exhibit nuclei. We can only say that we are acquainted with no fact whatever which speaks in favour of a further development, or of a cell division in the blood, but that everything points to the probability of a supply from without."--Virchow, *Cellular Pathology*, p. 223.

Frey, after a long disquisition on the blood, winds up as follows: "If we now ask, at the conclusion of this long inquiry into the nature of the blood, how much is known at the present day of the conditions during life of its two species of cells, we must allow that the results of all research, so far, are but very unsatisfactory."—*Histology and Histo-Chemistry of Man*, 1874.

Huxley says: "That the red corpuscles are in *some way or other* derived from the colourless corpuscles may be regarded as certain, but the steps of the process have not been made out with perfect certainty.' —*Elementary Physiology*, p. 62.

"McKendrick says: "As to the origin and end of the coloured corpuscles we are still in ignorance, nor can we trace precisely their relation to the colourless cells. Attempts have been made to show that they are probably the free nuclei of colourless cells, a supposition which has a few dubious facts to rest on, and which is not very probable."— *Outlines of Physiology*, 1878, p. 288.

Many statements of this kind might be brought together, but enough has been said to indicate the existing position of the subject.

[*] Recherches sur l'evolution des hématies dans le sang de l'homme et des vertébrés.—"Archives de Physiologie," MM. Brown-Séquard, Charcot, Vulpian.

[†] Ueber Knöchenmark und Blutbildung.

supposed, undergo atrophy, but escape bodily, while the coloured exterior protoplasm becomes gradually modelled into red biconcave discs.*

Simultaneously with the publication of Hayem's research in France (1878), I made known in this country my discovery of the existence of large numbers of *colourless and faintly-coloured biconcave discs* in the blood of mammals, a fact which had failed to be earlier observed, owing to the circumstance that the colour and refractive index of these bodies coincided *in the main* with that of the serum in which they lay. This discovery has since proved to be the key to unlock the mystery of the development of the blood, and has had the effect of harmonising all the various views which exist as to the extra-vascular sources of its corpuscular supply, for it can be demonstrated that all these sources yield to the blood as their most developed product *smooth free nuclei*, which

---

* Since the body of this work was printed, I have had the pleasure of perusing a paper in the January number (1882) of the *Archives de Physiologie*, by M. Malassez, in which he endeavours to show that the red discs which arise in the bone-marrow are produced by budding or gemmation of the coloured protoplasm of the red nucleated cells. Among a large number of photographs which I possess of bone-marrows of animals of various kinds and ages, I find a few examples only which could be interpreted to sustain such a view, and these do not belong to embryos or young animals in which, if anywhere, such a process might be expected to be proceeding at a rapid rate. The same thing occurs in the shed blood of the oviparous embryo, where the production of non-nucleated discs would be without meaning This inclines me, therefore, to hold still to the view of Neumann, somewhat modified, viz., that the entire cell, its nucleus, and protoplasm, undergo interstitial development, absorption, and condensation, and conversion, *in situ*, into a red disc. In such a view, the comparison of the relative volumes of the coloured protoplasm of the red nucleated cells and the red discs is obviously unimportant. Whatever may be the manner in which nucleated red cells become converted into red biconcave discs, the process (though persisting to an extent in some of the lower mammals) is essentially embryonic, and by no means represents the major process by which red discs are formed. In the bone-marrow, as elsewhere, naked nuclei (advanced lymph corpuscles) are transformed into colourless discs. In the embryo these nuclei sometimes become coloured before they leave the marrow, but in the adult they appear to pass into the blood in the colourless, invisible state.

appear in it at first as its *colourless disc*, and subsequently by the gradual attainment of colour become its red corpuscle. This is the case with the lymphatic glands, the spleen, the thymus, the thyroid, and the red bone marrow.*

---

* In a work recently issued by M. Hayem (1882), entitled "Leçons sur les modifications du sang," this author, whilst freely admitting the existence of the corpuscles above referred to, and the competency of my methods to display them, and proposing for them the new names of *achromacytes* and *chlorocytes*, adopts, to explain their presence, the exploded view that they represent *various stages of decolourised red discs*. M. Hayem appears to be very imperfectly informed as to the nature and extent of my work, and to have derived his knowledge from a hasty criticism in the "London Medical Record" (Jan. 1880) of the paper which I read to the Birmingham Philosophical Society, November 14th, 1878. He seems to have adopted, without personal examination, the conclusions of this writer, for they mutually labour under the erroneous impression that these corpuscles are the products of *compression and violence*, which discharges their hæmoglobin and renders them colourless. Hayem says : " Les hématies à une compression plus ou moins forte, un observateur anglais, Norris, a pu faire apparaître un nombre variable d'achromacytes, produits artificiels dans lesquels il a en la singulière idée de voir les formes primitives des globules rouges," page 287. In a second paper read to the Birmingham Philosophical Society in June, 1880, I have already replied fully to these objections, and have shown that *no compression whatever is necessary to observe these corpuscles, and also that red corpuscles cannot be decolourised by this means*. This paper, with additions, forms Section II. of this work. This notion of *compression* is an illusion which has arisen from the advice given in one of my experiments to firmly strap down the cover-glass upon the slide with its *convex* surface downwards, in order to produce the *barrier* of Newton's rings through which the liquor sanguinis filters off. It should be borne in mind that this is done before the blood is allowed to run in, and is only a matter of convenience, for many covers can be found with which strapping down is unnecessary, as they apply *themselves* properly to the slide ; besides, it is not necessary that the corpuscles should be squeezed or compressed *in the least ;* the glasses may be wide enough apart to allow them to fully retain their *biconcavity*, but not so separated as to permit them to slip over each other so as to become superimposed and to form a double layer. In this way we get the *colourless discs* surrounded by the *red ones*, and the former are seen because red ones do not lie above or below them to confer on them a borrowed tint. To see these *colourless and intermediately coloured discs* it is simply necessary to place a *small* drop of blood fresh from the finger upon a glass slide, and let down gently upon it a *thin* mica-cover about 1½ by ⅛ of an inch. What can be simpler? It differs only from the usual method of examining the blood in the size and flexibility of the cover, which conditions allow the blood to be seen in *single layer*. With a little

When Hewson threw out the suggestion that the " small
white solid particles " which he obtained from the glands were
the antecedents of the red corpuscles, he was partially right; but
when having observed in the blood the ordinary white corpuscle

practice their presence will be detected instantly the specimen is placed
under the microscope.   The same results are obtainable when the blood
has been kept at its normal temperature on the warm stage, or when
reduced to that of zero. Although in the sentence quoted Hayem commits
himself so distinctly to the statement that these are *artificial products*,
he nevertheless proceeds as follows : " Dans quelques les même, la présence
d'achromacytes dans des preparations faites avec le plus grand soin m'a
conduit à admettre que cette altération pouvait se produire dans le sang
en circulation, et avant l'action des agents extérieurs." Hayem further
suggests that moisture may play a part in these alterations.   This matter
was the first to engage our attention in 1877, when the corpuscle was
first discovered, for it was thought that the aqueous vapour arising from
the drop of blood *itself* might condense upon the slide or cover glass, and
lead to solution of the hæmoglobin.   This proved not to be the case.
The observation of Hayem that corpuscles can be decolourised by blowing
upon them with the mouth is an old and well-known one, which I practised
twenty years ago ; but it should be remembered that one of my methods is
the direct converse of this, as it consists in the simultaneous spreading and
rapid drying of the blood by means of *air blown from a large syringe through*
calcium chloride.   On the other hand red corpuscles decolourised by
moisture are entirely distinct from and unlike these colourless discs, and
would only be confounded with them by persons who had not made a
special point of studying them in contrast.

In Section II. of this work overwhelming evidence is furnished
that these corpuscles are not *decolourised discs*.   In Section III. the
several methods of staining with aniline blue are described, and when
so stained these corpuscles have quite a different appearance to
decolourised discs which have been similarly treated, the former being
large and plump, while the latter are thin and collapsed.   In Section IV.
a large amount of evidence is brought together tending to prove that
the so-called hæmatoblasts of Hayem are in reality but modified forms,
fragments, and granules of the slightly coloured discs of this series, of
the corpuscles, in fact, which he has designated *chlorocytes*.   In
Section V. proof is afforded that *colourless discs* (advanced lymph
corpuscles) are at definite intervals thrown into the blood from the
lymphatics and spleen, and in Section VI. the identity of these bodies
with the colourless discs of the blood is demonstrated.   In Section X. it
is shown that the phenomena of anæmia can only be explained by
considering the blood corpuscle to begin its career as *a colourless disc*
which acquires colour in a gradual manner in the blood.   In section XI. it is
shown that the *analogue* of these *colourless discs* is present in oviparous
blood as the *first stage in the blood* of the red nucleated corpuscle, and

he assumed its identity with the gland corpuscles, he committed the error which was repeated seventy-nine years later, when, in 1852, pathologists, having discovered that hypertrophy of the glands was associated with increase of white blood corpuscles, came to the conclusion that the function of the blood glands was to produce white blood corpuscles. I have shown that the function of the blood-glands is not to produce *leucocytes*, such as we see in the blood, but, on the contrary, *free nuclei*, which become its colourless discs. Our misfortune has consisted in failing to recognise that a progressive development continually goes on in the blood glands by which young and immature cells are deprived of their capsules and their nuclear discs set free. This process has a certain per centage of failure which may be regarded as normal to it, and which within certain limits is quite consistent with health. The number of leucocytes in any specimen of blood indicates the extent of this failure, because they represent the cells which,

---

that these *colourless ellipsoidal cells* are derived thus perfected from the lymphatics, bone-marrow, and spleen. As a matter of fact, my work may be regarded from beginning to end as a demonstration that these corpuscles are not decolourised discs, but *young discs which are passing from the colourless towards the fully coloured condition.*

As to the new names proposed, I am bound to say I regard them as both inaccurate and inadequate. In the first place, blood discs are not *cells;* and in the second, discs possessing every shade of colour exist between *achromacytes* and *chlorocytes* and these remain undesignated. The green corpuscles are those which I frequently refer to as the "green, lustrous, or diffused-edged corpuscles," and the most coloured of which photograph nearly as dark as the red discs (*vide* Photograph 59). These mark the limit of the *fugitive group of discs* which is constituted of those colourless and partially coloured discs which in their *degenerations* form fibrin and determine coagulation. The term *achromacyte* would be better applied to the *colourless stage* of the red nucleated corpuscle of the ovipara, because this is a true cell; but on the whole it seems to be undesirable to coin designations, however satisfactory they may seem, until we are agreed among ourselves upon the *facts*, for such terms are *catching*, become current, and, not being held tentatively, often in the end prove stumbling-blocks to scientific progress. No better example could be found than the confused use of the term *hæmatoblast*, which is now applied by different authors to widely-diverse elements.

having escaped capsular degeneration, become developed in the
large lymphatics and thoracic duct and in the blood into uni-
nuclear and multinuclear cells, and are there seen as the white
corpuscles. At this point we get a glimpse of the nature of the
defective function of the glands which gives rise to leukhæmia.
The production of leucocytes by the glands in the place of
nuclear discs is not to be regarded as an absolute failure of their
blood-forming function, but rather the reduction of it to a lower
and more sluggish type, for there are strong reasons for thinking
that the leucocytes of the blood shed their nuclei, and that these
pass through the stage of colourless discs and become ultimately
red corpuscles. To the uninuclear lymph cell, as found in
the blood-glands, I have given the name of the *primary lymph corpuscle*, and to its liberated nuclear disc that of the *advanced
lymph corpuscle*. Such few of the primary ones as escape
capsular degeneration become the *white corpuscles*, while the
main body which have undergone this change go to form
the *colourless discs* of the blood. In a restricted sense the
conjecture of Professor Gulliver was quite correct when he said:
" That the lymph globule is an immature cell which may change
in the blood, or even in the thoracic duct or lymphatic vessels,
into the larger and more perfect pale cell of the blood is very
probable." (Note cxlvi., p. 282, Hewson's Works.) This
of course could only happen with the *primary* lymph corpuscle.

It will naturally be anticipated that in the event of correct
views as to the mode of the development of the blood having been
obtained by these researches, new light will be thrown upon those
abnormal departures which we recognise as disease, and this has
proved to be the case, not only in leukhæmia, but also in the more
frequent derangement, anæmia. It transpires that the mode of
arrangement of the hæmoglobin in the individual corpuscles is
such as to be consistent only with a view which considers the
corpuscles to begin their career in a colourless state, and to
assume equal increments of hæmoglobin in equal times, we are
thus enabled to explain how it comes about that corpuscular
deficiency is associated with a greatly disproportionate loss of

hæmoglobin, and to show that this rests in a definite law, to which, with certain reservations, all cases of anæmia may be referred.

In the concluding sections I have sought to point out the part which the spleen and the bone-marrow play in the production of the blood of the ovipara, and have explained the close analogy which subsists between the development of mammalian and oviparous blood corpuscles.

SECTION I.

PART I.

On the existence in Mammalian Blood of a previously unknown Corpuscle, which explains the origin of the Red Disc and the formation of Fibrin.

THE discovery which forms the basis of this research was made during a protracted attempt to render photography available for the preservation of observations made with the *high powers* of the microscope.*

In order to carry out this object it was necessary to materially advance the existing state of micro-photography, and in the numerous preliminary experiments which this involved, a little blood drawn from the finger was commonly used as a test specimen. This led to the accumulation of a large number of photographs of human blood.

In some of these photographs corpuscles were observed which, although obviously in the same plane, were barely visible, and it was found that they could not be seen at all in the original specimens, however carefully looked for. Photography had therefore detected the existence of corpuscles which differed so little in refractive power and colour from the liquor sanguinis as to be invisible to the eye.

---

* It was desired to do this for the following reasons:—

1. It removes the doubt and distrust which are inseparable from hand drawing and engraving, however carefully and conscientiously performed, by furnishing illustrations which are as indisputable in the shape of evidence as the original specimens from which the photographs are taken.

2. It preserves perishable results so perfectly as to allow extensive comparisons of specimens to be made with a minimum amount of labour and fatigue.

3. Its well-known power to make apparent minute differences of *structure* and *colour*, which baffle the most trained eye, give it a claim to be regarded as a new and valuable *method of research* in Histology and Pathology.

Reflection on this curious fact gave rise to the opinion that possibly other corpuscles might exist, having precisely the same refractive index and actinic value as the liquor sanguinis, and that such would not only be invisible, but also incapable of being photographed.

Such a suspicion being aroused fresh specimens of blood were submitted in *extremely thin layers* to the most careful scrutiny, in the hope that such corpuscles might in some indirect manner betray their existence.

Under this *new condition* it was observed that the red corpuscles, as they moved about in the liquor sanguinis, occasionally became indented in outline, apparently in consequence of impinging against some unseen circular bodies less yielding than themselves.* By this observation the previous inference was materially strengthened, and a method was ultimately devised which rendered the presence of such corpuscles absolutely certain.

This method was based upon the idea that the spaces which these corpuscles occupied could not be otherwise filled, and hence if the liquor sanguinis could be drawn off in a great measure from *a very thin layer of blood* the red corpuscles would become applied to these bodies, and so form a new surrounding, which would render their presence obvious. This plan, when properly carried out in detail, as explained below under the heading of " The packing method," is most successful, for it displays both the size and number of these hidden corpuscles by a simple mechanical arrangement, which involves no alteration of the blood or addition of any foreign substance or re-agent to it, and allows of its examination within a few seconds of being shed.   By the

---

* Ordinarily these corpuscles are more limpid than the red ones, and are indented by them, as seen in Photographs 9 and 12, Plate III.: sometimes they even run in finger-like processes between the red corpuscles as a liquid might do. Examples of this may be seen in Photographs 15 and 16, Plate IV.   When they retain their circular outline and indent the red corpuscles it is probably owing to their having effected adhesion to the slide or cover glass.

methods which follow the existence of these corpuscles may be fully demonstrated.*

METHODS OF EXAMINATION.

BY PACKING.

From a number of thin cover-glasses a slightly convex one is selected by ascertaining that one surface (the convex) gives by reflected light a sharp image of the window bars, or gas-light, and the other a blurred and indistinct one. This cover is strapped, with its convex-surface downwards, upon a microscopic slide, so firmly as to produce a series of Newton's rings of an elongated form (*vide* Plate I., **Fig. 1**). Two objects are gained by this arrangement.

1. The glasses are everywhere in such close proximity as to admit the corpuscles flatways, and in single layer only.

2. In the part occupied by Newton's rings the proximity is much less than $\frac{1}{10,000}$th of an inch, and thus presents a *barrier* to the passage of the corpuscles, whilst allowing free passage to the liquor sanguinis. By this means the corpuscles are kept in the part marked A, whilst the liquor sanguinis is filtered off into B.

This packing of the red corpuscles in part A causes them to mould themselves around the invisible bodies present.

On Plate II., Photograph 3, is represented a specimen of blood from which the liquor sanguinis has been *partially* withdrawn by the method just described. The dark masses in this specimen consist in some cases of single deformed or indented red corpuscles, and in others of several such corpuscles fused together. Many of these masses will be seen to be bounded by *concave depressions* and if some of these concavities are

---

* The photographs illustrating Parts I., II., and III., Section I., are from specimens magnified 476 diameters, and those of Part IV. 500 diameters. The true relative sizes are therefore preserved. The actual size of any object may be obtained by measurement with the micrometer, Plate IX., Fig. 48, which is divided into 10 and 20,000ths of an inch.

carefully examined it will be observed that they are due to
the presence and influence of certain circular bodies, which
are either pressing into the red corpuscular masses, or forming
a basis around which these masses are applying and adapting
themselves. The latter view is probably the correct one,
inasmuch as the red corpuscular masses may often be seen
to swim freely about in the liquor sanguinis, still retaining the
curved outlines, and fitting into these curved outlines delicate
and nearly invisible corpuscles may be sometimes detected.
This being so, it is fair to consider that the concavities in
the cases where no corpuscles can be seen are also produced by
the presence of others which are still less visible. The
photograph just described represents, as stated, partial withdrawal of the liquor sanguinis. If this liquid is more perfectly
removed we get such specimens as are represented in Plate
II., Photographs 4 and 5. Here the whole of the dark part seen
between the invisible corpuscles is formed of red corpuscles
closely massed together, and by this means the spaces in which
the invisible corpuscles lie are rendered obvious, and although
the blood may be said to swarm with these corpuscles they
escape observation, because their *colour* and *refractive index*
coincide exactly with that of the liquor sanguinis.

In Plate II., Photograph 5, the packing is less complete,
and the individual red corpuscles forming the dark background
can be detected.

BY ALTERING THE REFRACTIVE INDEX OF THE LIQUOR SANGUINIS.

If we place upon the tip of the finger a minute drop of
saturated solution of salt, and prick through it, so that the
blood may flow directly into the saline solution, the refractive
power of the liquor sanguinis is modified, and it is found that
if we run this mixture of salt and blood between glasses prepared according to the packing method before described, we can
then see the *outlines* of these colourless, circular bodies where
they lie in contact with each other, and we find that many of
the clear spaces which have hitherto been supposed to consist

of liquor sanguinis only, are really occupied by these colourless
discs, as seen in Plate II., Photograph 6.   After a time specimens thus prepared become tinted with hæmoglobin, and this
stains the edges of the colourless discs, and renders them still
more apparent, as in Plate III., Photograph 7.

BY ISOLATION.

From what has been said, it is obvious that these corpuscles
cannot be properly examined so long as they remain submerged
in the normal liquor sanguinis ; it is therefore necessary to
devise some means by which they can be freed from this liquid,
and at the same time isolated as much as possible from other
corpuscles.   Numerous observations having pointed to the fact
that the adhesiveness of the blood corpuscles to foreign substances, *e.g.*, glass, was inversely as their degree of colour, this
was taken advantage of as a means of separating them from
the liquor sanguinis and from each other.

The kind and character of the corpuscles withdrawn will
in great measure depend upon the length of time the blood is in
contact with the glass.

Capillarity is the principal means that has been employed
to get rid of the main body of the blood.   Thus cover-glasses of
various and often considerable sizes were used, and these, in
some cases, were firmly strapped down *at both ends,* with good
adhesive plaster, to a slide having a hole drilled in the point of
intersection of its diagonals.   The blood could then be introduced at the central hole, or at the circumference of the cover-
glass near to the hole ; then, by carefully springing up the cover-
glass, by introducing a tightly-fitting wooden plug into the hole,
the blood can be made to recede by capillarity into another part
of the arrangement, leaving the adhering corpuscles free for
examination.   Instead of the plug, a fine screw working in a bush
was sometimes used, for by this means the blood can be removed
more gently, and also be allowed to return again over the
adhering corpuscles, permitting an opinion to be formed as to
their degree of visibility, etc., when so submerged.   This plan
has the decided advantage of being perfectly under control. and

therefore more gentle and gradual in its operation. (*Vide* Plate I., Fig. 2.)

On other occasions cover-glasses were strapped down at *one end only*, forming a kind of hinge, which permitted them to be raised by the screw, or otherwise, from the opposite end, and the excess of blood would then flow by capillarity to the other, from whence it could be removed by blotting paper, and the adhering corpuscles retained for examination. (*Vide* Plate I., Fig. 3.)

Another method of getting rid of the main quantity of blood was, to insert a straw or quill into a hole drilled in the glass slide, and either to forcibly blow the redundant blood into another part of the covered space, or suck it from the circumference towards the centre. In some cases, when it was sought to isolate corpuscles that had as yet attained little or no colour, and which are apt to spread themselves down almost immediately on the glass and become lost to observation, air that had first passed through dilute solution of osmic acid, and subsequently through chloride of calcium was used. When displacement of the blood was effected by blowing, the arrangement depicted in Plate I., Fig. 4, was adopted, but when by suction, the rubber ball was replaced by the mouth.

When such methods are employed in their simplest form groups of corpuscles are found to be adhering to the cover glass and slide, of the character seen in Plate III., Photograph 8. These mosaic groups appear to form while the corpuscles are still submerged in the liquor sanguinis. They result from cohesive attraction, operating in a progressive manner, as explained by the Author in his papers on the formation of rouleaux and on the passage of the corpuscles through the vessel walls.*

---

* A consideration of the causes of various phenomena of attraction and adhesion in solid bodies, films, vesicles, liquid globules, and blood corpuscles.—Proceedings of the Royal Society, London, 1862. Aggregation of the blood corpuscles both within and without the vessels (rouleaux).—Proceedings of Royal Society, London, 1869. Principles concerned in the extrusion (without rupture of the vessels) of the morphological elements of the blood (so-called emigration of corpuscles). —Proceedings of Royal Society, London, 1871; Transactions of St. Andrews Graduates' Association, 1871.

The action may readily be imitated by smearing a little soap solution upon a glass plate, and producing minute bubbles with the aid of a capillary tube. The bubbles will be seen to arrange themselves in groups, with flattened facets, just as the corpuscles do. In the case of the latter the action is due to direct cohesion, while in the former it is owing to capillarity (double or indirect cohesion). It will be noticed that when one of the minute bubbles burst the others are immediately drawn up, and the whole group undergoes rearrangement, so that no vacant space is allowed to remain. It has been observed that a precisely similar thing occurs with these blood groups, and that the invisible corpuscle is the one which by falling into a liquid state, and spreading itself on the glass slide, yields up its place to the more stable ones. This sinking down of some of the more delicate corpuscles and the rearrangement of the more stable ones, occurs with such rapidity, that it is only by having the groups under the eye at the moment that the liquor sanguinis is removed from over and around them, that it can be seen. A moment later all is quiet, and there is nothing to excite the least suspicion that such an important modification has occurred.

The first corpuscles to attach themselves to foreign bodies such as glass, are the uncoloured, and very slightly coloured, smooth corpuscles, the invisible corpuscles of the blood. But these are easily overlooked; for they are so delicate and fragile, that when they do not melt down, so to speak, they almost invariably break up into molecules or spheroidal granules, however gently the main quantity of blood may be withdrawn by capillarity from off and around them, *i.e.*, however carefully the glasses may be separated from each other. Another mode in which they frequently elude observation is by adhering and fusing together so as to form delicate films upon the slide and cover-glass, and they often seem in this way to furnish a basis of adhesion for the more coloured corpuscles; for mosaic groups of these are constantly found to have a delicate layer formed of the uncoloured corpuscles beneath and around them.

On Plate III., Photographs 9, 10, 11, 12, and 13, show groups of corpuscles which have been withdrawn from the blood by means of their adhesiveness to solids. Among them, the corpuscles in question may be seen. In Photograph 9, five such may be counted, and in Photograph 10 there are three which are blending into a smooth mass, and five others which possess some colour. Photograph 11 is a group in which four corpuscles have already blended into a smooth mass, and in the upper part of the specimen a similar mass shows a tendency to granulation.

Photograph 12 shows the new corpuscle in an isolated state, and also masses resulting from their fusion, some of which have become granular.

Photograph 13 shows numerous isolated invisible corpuscles, some of which exhibit faint traces of colour. All these specimens were obtained by the plan of separating the glasses gently by the use of the fine screw, as depicted in Plate I., Fig. 3.

BY OSMIC ACID VAPOUR AND COLD.

As in some of the examples yielded by the previous methods there appeared to be indications that the *normal* form of the transparent colourless corpuscles was that of the un-altered red corpuscles, viz., a biconcave disc, it was decided to attempt the preservation of their true form by separating the glasses over a shallow pan filled with the vapour of osmic acid, it being well known, as pointed out by Schultze, that this vapour possessed the property of preventing change in the form of the red corpuscle. The series of Photographs 14, 15, and 16 on Plate IV. were obtained by this method, and are remarkable groups for displaying the fact that corpuscles exist of every gradation of tint, from those which are perfectly colourless to the fully coloured red disc. In Photograph 16 we have also evidence that the corpuscles which are freest from colour possess in their original state a biconcave form. The most decisive results, however, as to the biconcave form of the new corpuscle have been obtained by the use of cold and osmic acid vapour in conjunction. The slide with the hinged cover glass (*vide* Plate I., Fig. 3) is wrapped up in blotting paper, and placed between two blocks of ice, and allowed to remain till it has attained the ice cold temperature. It is then unwrapped, and the finger having been previously pricked, the blood is allowed to run in under the cover-glass before any condensation of moisture upon the slide has had time to take place. The slide is immediately placed face downwards over the osmic acid pan, and the glasses gently separated by the screw. The form of the new corpuscle can thus be perfectly preserved, as seen in Plate IV., Photographs 17 and 18.

BY ALCOHOL AND SOLUTION OF OSMIC ACID.

The use of the vapour of osmic acid having proved so valuable as a coagulant in securing the presence of, and arresting the changes in, the invisible corpuscle, other vapours and substances were used, and among the rest, alcohol. Some very successful specimens were obtained by means of this agent. The method consisted in the use of one of the hinged coverglasses before referred to. Absolute alcohol was introduced between the cover-glass and slide till the space was filled, then a drop of blood from the end of the finger being placed at the edge of the cover-glass readily finds its way between the glasses, and mixes with the alcohol. After a short time, the glasses being separated by means of a screw, groups of corpuscles are found attached to their surfaces, and in and around these groups numbers of the invisible corpuscles were to be seen. Such groups are seen in Plate V., Photograph 19. A keen sight may detect in every part of the background of the photographs of such specimens faint dark lines : these indicate the existence of a layer of phantom corpuscles which have become fused, and have spread themselves out in a delicate fibrin-like layer upon the glass. By using a two per cent. solution of osmic acid in the same manner as alcohol in the previous case, they may be much more fully preserved, and then by protracted staining they can be brought into view, as in plate V., Photograph 20. This specimen underwent prolonged staining with alcoholic solution of aniline brown, and a complete background of corpuscles which were previously invisible was thus brought into view.   The corpuscles themselves have not taken the stain, but it appears to have penetrated into the interstices between them.   Plate V., Photograph 21, is from a similar specimen, but shows, in addition, that prior to the arresting action of the osmic acid many of the invisible corpuscles had become already fused into liquid masses, which appear to adhere to and surround the more coloured corpuscles. In many cases the invisible corpuscles are still sufficiently distinct and intact to show that these " fibrin or plasmine pools " originate, from their fusion.   There is reason to think that every part of the background of this specimen is covered with liquid of corpuscular origin.

Plate V., Photograph 22, shows invisible and slightly coloured corpuscles in the act of spreading, fusion, and disintegration.   It represents a still more advanced stage in plasmine formation.   It will be noted that while the invisible corpuscles have refused the aniline brown the partially coloured ones among them have become deeply stained by it.

BY SPREADING AND DRYING.

Mammalian blood can be readily spread out upon a glass slide, so that the corpuscles shall exist in a single layer only. The best method of doing this is to take a strip of flat glass about three inches in length, and a quarter of an inch in width.   The drop of blood being placed between the strip and the slide, at one end of the latter, the former is smoothly and slowly moved along the glass to the opposite end.   It is interesting to observe the difference between the specimens of blood simply spread by the above method and allowed to dry spontaneously, and others in which the drying is accomplished artificially, as the spreading proceeds, by blowing large quantities of air upon it from a small blast-fan, or hand bellows.   In the former the spaces between the corpuscles which represent the dried liquor sanguinis are clear and transparent, while in the latter case they are comparatively opaque and turbid, owing to the presence of the invisible and faintly coloured corpuscles. Some distortion of the corpuscles is produced by this method, but it has the advantage of allowing their relative number to be approximately estimated.

Plate V., Photograph 23, shows the appearance when drying is allowed to take place spontaneously, and Photograph 24 when it is expedited by blowing. In the latter case the fugitive corpuscles are prevented from melting down into a film of fibrin upon the surface of the slide.

There are two general principles by which these corpuscles may be made visible while floating in the liquid of the blood. When we reflect that they have the same colour as the liquor sanguinis, it is obvious that substances having the power of tinting this liquid, but incapable of staining the corpuscles, should render the latter visible. This is actually the case. The second principle is that of staining the corpuscles themselves *more deeply* than the liquor sanguinis. The corpuscles must, therefore, in this case have an affinity for the stain.

BY TINTING THE LIQUOR SANGUINIS.

Three-quarter per cent. solutions of sodium chloride charged with such substances as hæmoglobin, caramel, and saffron, have been found most successful. In all these cases the invisible corpuscles appear, and remain as delicate, colourless bodies, contrasting strongly with the ordinary red discs.

In this method, and also in that of actual staining, there should be no depth of liquid above or below the corpuscles, as this entirely obscures them. They are best seen when the glasses are so closely approximated as to actually compress them, and to increase their diameter when lying flatways, and as in such cases the layer of stained liquor sanguinis is less than 1-10,000th of an inch in thickness, the staining fluid requires to have in the first instance the utmost intensity : if this is not the case, sufficient difference fails to be created. The slides should be prepared as for " the packing method."

In obtaining the specimens of blood for examination the following method should be adopted :—Place upon the

end of the finger a small drop of the staining fluid, and
with a needle prick the finger through this drop, so that the
blood may, when the finger is squeezed, flow directly into the
liquid, which has the double property of both *preserving and
tinting or staining*, as the case may be. After mixing well with
the needle on the end of the finger, the blood may be allowed
to flow by capillarity between the cover-glass and slide for
examination.

BY STAINING THE INVISIBLE CORPUSCLE.

The utmost caution is of course demanded in drawing
conclusions after the addition of reagents of any kind to the
blood. Such a method is always open to the objection that the
reagent may produce that which is simply sought to be
distinguished. This objection is met by the adoption of
methods which show unequivocally that the corpuscles which
take the stain are the self-same corpuscles as those
which occupy the spaces that appear in the 'packing
method.' Having once satisfied ourselves that such corpuscles—and such alone—take the stain, while the ordinary
red discs refuse it, we have no difficulty in displaying and
distinguishing the *fugitive group* of discs, and in demonstrating
that essential chemical differences exist between these and the
hæmoglobin group, but that in this respect they gradually merge
into each other.

When a staining fluid is added to blood its first action is
to bring into view the invisible corpuscles as delicate white
bodies, which contrast strongly with the ordinary biconcave
discs; these, then, become rapidly stained in the inverse order of
their colour till a limit is reached, when no further staining
occurs. As the corpuscles take up the dye they of course become
again invisible, because they assume the same colour as the
tinted liquor sanguinis. After some time has elapsed they,
however, reappear, having acquired a stronger tint than the
liquid which surrounds them.

The subject of staining is one of such great importance
both as to differentiating these corpuscles from the red ones,

and also in tracing their origin, that it has been deemed desirable to devote Section III. specially to its consideration.

From the observations I have here recorded I consider that two conclusions are justifiable,—1st, That there exist in the blood of mammalia, in addition to the well-known red and white corpuscles (*vide* Plate II., Photographs 1 and 2), colourless, transparent, biconcave discs of the same size as the red ones; 2nd, That between these two kinds of biconcave discs others are demonstrable, having every *intermediate* gradation of colour.

The *origin* of these new corpuscles is treated of in the second part of this Section.

Part II.

On the Origin of the Colourless Biconcave Discs of Mammalian Blood.

The existence of these *colourless biconcave discs* having been demonstrated by a variety of methods, and their number shown to be considerable, it becomes important to ascertain their source.

Three conjectures appear to be admissible. 1st, They may be regarded as red corpuscles which have become decolourised during the mere act of shedding the blood. 2nd, They may be considered as representing stages in the dissolution of the red disc. 3rd, They may be biconcave corpuscles which are destined by acquiring colour to become converted into the red discs.

Against the view that they are corpuscles which have lost their hæmoglobin in the brief interval (a few seconds) between the shedding of the blood and its examination, it may be urged,— 1st, That they are obtained in greatest perfection when those measures are adopted which tend to preserve the blood from change, *i.e.*, cold or osmic acid. 2nd, They may be seen immediately the preparation is made (packing method) *and* before the liquor sanguinis has become stained, and they do not increase in number as time elapses. 3rd, Assuming the red corpuscles to lose hæmoglobin, the loss must occur in such a manner as to furnish corpuscles exhibiting all gradations between *a colourless* and *a full red* biconcave disc. If, therefore, we start with the idea that the corpuscles originally are all coloured, their decolourisation *in this graduated manner* would of itself indicate a difference in nature. 4th, If a three-quarter per cent. solution of salt be saturated with hæmoglobin, and a drop of this be placed upon the end of the finger, and the latter be pricked through the drop so that the blood may come at once into contact with fluid *saturated with hæmoglobin*, the colourless discs are still present as usual. It is obvious that these are not conditions favourable to the yielding up of the colouring matter. 5th, When first brought into view by the method of altering the refractive index of the liquor sanguinis, many of these corpuscles are of a pure white colour, but they gradually become stained by the hæmoglobin discharged by the red corpuscles. (*Vide* Plate II., Photograph 6, and Plate III., Photograph 7.) It is impossible to suppose that they first yield up the *whole* of their colouring matter, and then subsequently take it up again. 6th, Finally, the general behaviour of these corpuscles after the blood is shed, their tendency to break up into granules, to lay themselves down as delicate films, to form networks ; in a word, their fibrin-forming property is totally opposed to the conception that they were once red discs.*

The second conjecture, that they are red discs which have lost their colour as a preliminary step towards dissolution, has no facts to lend it support, while it is easy to show the existence in the blood of dark red granules, to which it is difficult to attach a meaning, unless we regard them as red corpuscles undergoing disintegration. These often occur in masses as in Plate VI., Photograph 26 ; but the same disintegrating action may be traced in single corpuscles.

The third supposition, that they may be biconcave discs which are gradually assuming colour, and which, therefore, exhibit every gradation of tint during their transition stages,

---

* Vide Section II., Page 41.

necessarily involves the consideration of the existence of *a source*, internal or external, from which the blood can be continually supplied with the large numbers of colourless biconcave discs which are seen to be present in it. We are thus naturally brought to the consideration of the morphological elements found in and derived from the lymphatic glands and spleen.

LYMPH AND SPLENIC CORPUSCLES.

It is desirable, in the first place, to state that all observations made on the corpuscles of the glands and spleen are untrustworthy and comparatively worthless unless made almost immediately after the removal of these organs from the body, and while they are still *warm* and *fresh* ; for typical and unchanged examples of splenic and lymph corpuscles can only be obtained directly after death, or by the method of immediate and rapid freezing.

This fact has not to my knowledge been previously urged, and it will, I believe, account for many of the contradictory statements made by different investigators respecting the nature of the glandular and splenic pulp.

The changes of which these bodies are susceptible may be described as Enlargement, Aggregation, Fibrillation, Spreading, or Lamination.

ENLARGEMENT.—Soon after death the lymph and splenic corpuscles undergo change both in *form* and *size*.   The average diameter of the lymph and splenic corpuscles of the bullock, pig, and sheep, when obtained from warm glands or spleen, or glands and pieces of spleen which have been frozen, is 4-20,000ths of an inch.   These measurements were always made upon examples which were free from the disposition to aggregate or spread— *vide* Plate VI., Photographs 27 (lymph) and 28 (splenic).   As the organs cool (especially is this the case with the spleen), the corpuscles undergo change of form and increase in size, attaining diameters of 5, 6, 7, and 8-20,000ths of an inch, the average being about 6-20,000ths of an inch.   What this spontaneous increase of size is due to is not very apparent, but we know that any alteration made in the environment of these bodies is attended with important changes of this kind. If, for example, we add one per cent. solution of sodium chloride to the fresh splenic corpuscles of the pig they swell up, and appear to gain greater smoothness, and to display considerable body.

The behaviour of the lymph corpuscles when treated with water is also remarkable. They often swell up into globular masses, having diameters of 15-20,000ths of an inch. This property was well known to Hewson. Plate VI., Photograph 29, shows spontaneous enlargement; Plate VI., Photograph 30, after addition of one per cent. saline solution; and Plate VII., Photograph 31, after addition of water. The highly swollen corpuscles in this specimen were so faint that it was necessary to stain them, hence their dark colour in the photograph.

Aggregation.—These corpuscles readily coalesce with each other, forming larger masses. These may be distinguished from merely swollen states by their superior density or opacity ; sometimes they coalesce to such an extent as to form a thick liquid, which will stream down the slide. Examples of their aggregated states may be seen in Plate VII., Photographs 32 and 33.

Fibrillation.—The corpuscles frequently adhere to each other and become extended into fibres, as in Plate VII., Photograph 34.

Spreading and Lamination.—Some of the corpuscles frequently spread, coalesce, and melt down upon the glass into a continuous soft film, to which others, less susceptible to this action, adhere. Plate VII., Photograph 35, represents this condition.

The singular manner in which these simple structures behave, both in relation to each other and to the surfaces with which they are necessarily brought into contact, when submitted to microscopic examination, indicates that they are very little removed in their essential nature from liquids. They appear, in fact, to occupy an intermediate position between limpid liquids and pasty solids, and may, according to their surroundings, exhibit phenomena which belong to either one or the other condition, and hence they sometimes lose all differentiation of parts and act precisely like liquid substances.   It is this behaviour which renders it so difficult to arrive at their true form and constitution.

The general liquid character of these corpuscles may be judged of by the disposition which they display to coalesce, to spread, or lay themselves down into soft films, which sometimes run like a treacley liquid.   (*Vide* Plate VIII., Photograph 37.)

Varying degrees of liquidity, however, exist among them, for some have a much greater power of retaining their shape than others.   The same variation in behaviour is also seen with reagents.

The corpuscles most prone to change are the oldest or most developed.   This, it will be noted, is precisely opposite to what occurs with the blood corpuscles.

The lymph and splenic corpuscles are constantly spoken of as *globules*, and are generally held to be *spherical* in form.   This has arisen from the difficulty connected with ascertaining their real form, and from their proneness like the blood disc to change and assume the globular shape.   In their normal state they are *discoid* bodies, with numerous slight irregular depressions upon their plane surfaces, which give to them their granular or corrugated appearance.   Having a specific gravity slightly less than the liquid in which they exist they invariably float, as cork discs would do, with their plane surfaces uppermost.

Occasionally, by dexterous pressure upon the cover-glass, they may be made to turn over, and their true form is revealed. This can rarely be done.   Their *discoid* form can, however, be determined by indirect methods, one of which is as follows :— Ascertain by the micrometer the average diameter of a number of these bodies, to be as usual from 4 to 5·20,000ths of an inch, and then take a slide arranged as in the method for packing, which will only allow blood corpuscles to pass in *flatways*.   If the lymph corpuscles pass between these glasses without material increase of diameter they must be regarded as discs.

They may also be rendered irregularly spherical by means of a two per cent. solution of salt. They then become smaller in diameter, and roll over easily, and appear very like blood corpuscles which have taken on irregular cup-shaped forms. (*Vide* Plate VIII., Photograph 38.)

As before stated, these corpuscles when unaltered are *discoid bodies*, with numerous slight irregular depressions, which give a granulated appearance to their plane surfaces. These depressions appear to be the commencement of the process or operation by which the disc is eventually rendered biconcave; for if the specimens of lymph are preserved for a time under sealed cover glasses they may be seen to undergo conversion into biconcave discs, and the transition stages of the process may be watched.

This property of becoming biconcave appears to belong to the corpuscle as a *physical substance*, rather than as an *anatomical structure*, because part of the corpuscle will become smooth and regularly biconcave, while the other part still retains the character and appearance of a lymph corpuscle; hence we have the instructive spectacle of one part of the same corpuscle having the character of a colourless biconcave disc, and the other that of a comparatively unaltered lymph corpuscle. (*Vide* Plate VIII., Photograph 39.)

The fact, too, that biconcave discs can be artificially produced by placing certain colloid substances in the proper relation to each other is a further argument in favour of this view. (*Vide* Plate **VIII.**, Photograph 40.)*

This capacity of the lymph and splenic discs to undergo conversion into smooth colourless biconcave discs seems to point definitely to the *source* of the *colourless biconcave discs* found in the blood. The reader will, however, find this aspect of the subject treated in much greater detail · in Sections V. and VII.

---

* For further information on the causes of biconcavity the reader is referred to Part IV. of this Section.

Comparison of the Lymph, Splenic, and White Blood
Corpuscle.—A great deal of conflicting statement exists as to
the relation which these bodies bear to each other.

Many writers speak very loosely of the *identity* of the
lymph and splenic corpuscles with the white blood corpuscle.
Others are more critical, and claim that important distinctions
exist between these bodies. Professor Gulliver is among the
latter, and says: "The globules of the chyle, of the thymus
fluid, and of lymph, are *smaller*, and differ in structure from the
pale globules of the blood. In these last there are two, three,
or four nuclei, easily seen when the envelope is made more or
less transparent or invisible by acetic, sulphurous, citric, or
tartaric acid. But the globules of chyle, of lymph, and of
the thymus fluid, like the nuclei of the red corpuscles of the
blood,* are only rendered more distinct and slightly smaller by
any of these acids, so that the central part presents no regular
nuclei, or divided nucleus, such as are contained in the pale
globules of the blood. In short, these last named globules
have the characters of perfect elementary cells, while the
former globules resemble and probably are nuclei or immature
cells."

Gray also speaks of the splenic corpuscle as a nucleus.

No one can compare Photographs 27 and 28, Plate VI.,
with Photograph 2, Plate II., without being astonished that
bodies so essentially unlike in *size, form*, and *colour*, should have
been so confounded.

1st. The lymph and splenic corpuscles are *discs*, having a
diameter varying from 4-20,000ths to 5-20,000ths, and a thickness of about 1-10,000th of an inch.

2nd. The white blood corpuscles are *spheres*, having a
diameter varying from 5 to 10-20,000ths of an inch.

3rd. The lymph and splenic corpuscles consist of small uni-
nuclear corpuscles, and of free disc shaped nuclei, and it is the
latter only which enter the blood to any extent in health.

---

* The nucleated corpuscle of the lower vertebrates is here referred to.

4th. The white blood corpuscle has from *one to four* or even *five nuclei.*

5th. The lymph and splenic discs can become spherical; but they then have a still *smaller* diameter.

6th. The white corpuscle is *whiter* than the Liquor San-guinis,* while the lymph disc at the time it enters the blood is of the same tint as this liquid.

The relation which subsists among these bodies is much more exhaustively discussed in Sections V., VI., and VII.

PART III.

The Granule Sphere (Körnerkugeln).

IN *Pflüger's Archives* for November, 1875, will be found a paper by Alexander Schmidt, in which publicity is given to certain observations made on the blood by Semmer. This investigator examined microscopically the plasma of horses' blood which had been prevented from coagulating by means of cold, and arrived at the following conclusions :—

1st. That there exist in the plasma of horses' blood which has been allowed to subside at an ice-cold temperature, yellow and red granule balls or corpuscles (rothe Körnerkugeln) which are considerably larger than the ordinary colourless corpuscle of the blood.

2nd. When kept for a few hours " these corpuscles dis-integrate into white granular heaps " (farblose Körnerhaufen).

3rd. When these granule balls are treated with CO_2 the red granules partly disappear, and a colourless nucleus becomes visible.

Under this treatment, when complete, the red granules entirely disappear, and a white or yellow nucleus is visible.

4th. Water acidulated with acetic acid causes the nucleus sometimes to be extruded.

---

* This effect seems to be mainly due to the different relations which a granulated and a smooth surface bear to light. It may be illustrated by ground and smooth glass.

5th. Acetic acid causes the red granules to agglomerate in the centre, whilst a homogeneous white border becomes visible.

6th. The red granules have disappeared, and a yellow nucleus lies in a finely granulated colourless mass, clearly distinguished from the homogeneous border.

7th. An analogous action occurs with the frog's blood corpuscle. Under the action of acetic acid the nucleus becomes completely covered with yellow granules. Later on the smooth and yellow nucleus lies within the colourless cell wall.

It is suggested by Semmer that these granule balls or spheres occupy an *intermediate position* between the ordinary white corpuscle and the red disc, that they are, in fact, the transition stages in the development of the red disc.

It is assumed, therefore, that they exist in *perfectly fresh* blood, *i.e.*, in blood circulating in the vessels, but disappear immediately when the blood is shed, going in some way to form fibrin.

It is further assumed that *shed blood* which has been prevented from coagulating by cold may be regarded as *fresh blood*, and may be held to represent the conditions which obtain in the vessels.

In Semmer's method some time is allowed for the red corpuscles to sink, so that the upper stratum of plasma containing a few corpuscles only may be taken for examination.

The question arises, can blood thus obtained and treated be taken as normal ? Is the fact that *coagulation* is prevented, to be taken as proof that *no changes* occur in the corpuscular elements ?

It is obvious that blood may be examined *earlier* than by the method of Semmer, with the same precautions, and under still better conditions. This may be done by fastening a slide with attached cover to a small block of ice, and then pricking the ear of the horse, or the human finger, bringing the blood at once *in a thin layer* under the influence of cold. It may then be examined on the " cold stage "* of the microscope.

---

* The ordinary warm stage traversed by ice-cold water.

In *horses'* blood thus obtained there is never to be seen anything besides the ordinary red and white corpuscles, unless the glasses are prepared according to the 'packing method,' when the invisible and slightly coloured corpuscles are also rendered obvious.

When *human* blood is examined on the " cold stage " of the microscope, the necessary precautions being taken to bring down the temperature of the slide and cover to zero before introducing the blood, I find that the transition corpuscles are exceedingly well preserved ; but no bodies can be seen of the nature of granule balls.

Although the younger corpuscles are by these methods preserved from melting down they have a great tendency imparted to them by the cold to break up into granules, and it is of these aggregated granules that the granule balls are composed.

Plate IV., Photographs 17 and 18, represent specimens of human blood which were subjected to the ice-cold temperature immediately the blood left the vessels. They furnish very good examples of the breaking up of young coloured corpuscles into large granules.

The nucleated appearances observed in these granulated spheres, when they are treated with acids, have a peculiar interest in connection with the physical and chemical con-stituents of the biconcave discs.

These appearances may be studied in Plate IX., Photograph 47, in which the effects produced by acid on biconcave discs possessing various degrees of colour may be seen. That these corpuscles appear to be nucleated cannot be doubted, yet there are other ways of viewing such results.

The facts are briefly these :—A *colourless disc* acquires colour *gradually* by the transformation of some of its substance into hæmoglobin. Till such time as it is wholly converted into hæmoglobin (assuming such to be possible) it will consist of two substances ; the substance of which the colourless disc is com-posed plus hæmoglobin ; the younger the corpuscle the greater the amount of *unaltered white substance* present. The hæmoglobin may either be superficial or diffused throughout the mass.

When the corpuscle is only in part converted into hæmoglobin, the white and yellow constituents may become separated from each other, and thus give rise to various appearances; among others, to that of a nucleated cell, as seen in Photograph 47.

It is in this way that the so-called nucleated cell obtained by the action of acid on the granule spheres, described by Semmer, is produced.

When blood is allowed to coagulate the granule balls of Semmer mainly disappear, being simply the more coloured corpuscles of the *fugitive group.* They melt down into fibrin without undergoing granulation, but when by the use of cold this is prevented, they undergo granulation and show themselves as coloured granule spheres. When the corpuscles first break up these granules are coloured, but they subsequently, *i.e.*, in "a few hours," give up their colour to the *liquor sanguinis* and appear white.

It is a curious fact that the separation of these granules does not reveal any nucleus, which it should do if such a body exists as a separate and distinct structure.

The nucleated appearance is in *all cases* due to reagents, and to reagents which I have shown to be capable of producing the same appearance in the *ordinary red corpuscle.*

The so-called nucleus thus brought into view is stated by Semmer to be sometimes of a yellow tint, and to become extruded ; and it is inferred that this *extruded tinted mass* is the initial stage of the red corpuscle. No explanation as to how or when it acquires *biconcavity* is afforded.

The yellow tint is a mere stain imparted by hæmatin, which has become dissolved in the acid, and is not a genuine development of colouring matter. This is proved by one of Semmer's own experiments, in which the frog's corpuscle being treated with acids the nucleus becomes tinted. The two cases are analogous; but while one might be made to support a theory, the other would be without physiological meaning.

From what has been said it is obvious that the accurate observations made by Semmer *when properly interpreted* lend no support whatever to the theory which regards the ordinary white corpuscle of the blood as the precursor of the red disc.

SUMMARY.

The views of the development of the blood indicated by the present research may be thus briefly summarised :—

1.—There exist in mammalian blood numerous corpuscles which are incapable of being seen by the microscope, not because of their ʼminuteness, but owing to the fact that they have the same *refractive index* and *colour* as the *liquor sanguinis* in which they are submerged. When brought into view and carefully examined by suitable methods they prove to be *colourless biconcave discs*, and between these and the *red discs*, biconcave corpuscles possessing *every gradation of tint* can be detected.

When the fully developed elements of the lymphatic glands, spleen, etc., are examined with the same precautions for the preservation of their *size* and *form* as are necessary for the blood corpuscles, instead of being found to be identical with the ordinary white blood corpuscle, they prove to be *discs of the same size as the red corpuscle which are gradually becoming biconcave.*

3.—As we know that these delicate and already scarcely visible bodies are poured into the blood in large numbers at the subclavian and splenic veins and are henceforth *lost to view*, it is not unreasonable to conclude that they are represented in the blood by the *colourless biconcave discs* before referred to, and that they become completely *invisible* by transference to a liquid having the same *colour* and *refractive index* as themselves, reappearing gradually as they gain sufficient hæmoglobin to contrast with the *liquor sanguinis*, and before this occurs they have fully perfected their biconcave form.

Part IV.

On the Cause of the Biconcave Form of the Mammalian Blood-corpuscle.

In a recent contribution on the blood,[*] Prof. Rindfleisch has endeavoured to account for the biconcavity of the red corpuscles by mechanical considerations. He says: "The biconcave shape of the red blood corpuscle is dependent on the one hand on the original bell-shaped form[†], and on the other on the rolling brought about by the blood current. This rolling is not at all a simple onward movement in one direction, but at the same time a mechanical modelling of the blood corpuscles by contact with firm or resistant bodies, which have the same influence in modelling their form as the grinding and mechanical contact to which stones are subjected has upon them while lying exposed in running water. No biologist can have overlooked the resemblance between so-called 'roll stones' and red blood corpuscles, those remarkable, regularly formed discs, with which, as a boy, he has tested the elasticity of the water. If bell-shaped bodies are made of putty, which must be very stiff, and they are rolled about in a bottle of water, then flat discs are formed, or (and this may appear more striking when we recollect the mechanical treatment to which a blood corpuscle is subjected in the stream of blood, through the simultaneous rolling against the wall of the blood vessel, and compression through impact in front) if we roll the bell-shaped form, at first gently, between the hollowed palms, and then compress the sides, a hollow ball is formed by the rolling, and the well-known meniscus by the compression. This treatment is not identical with the very simple formation of a biconcave form by compression of a ball of clay between the thumb and finger, for in the former experiment the edges of the bell first join

---

[*] "Ueber Knöchenmark und Blutbildung." Von Dr. G. E. Rindfleisch, Professor in Würzburg.

[†] For the meaning of the expression *original bell-shaped form* the reader is referred to Section VIII., "On the rôle of the red bone marrow in blood formation."

like seams. (The shape of this seam reminds one strikingly
of the appearance which the blood corpuscles exhibit when
treated with a solution of alum, when it is on the point of
taking its old bell-shape.) The biconcave meniscus is formed
almost of itself, without much pressure, if care is taken that
the air can escape freely from its interior. A thick-sided ball
also forms on compression a biconcave meniscus, without any
important derangement of even the smallest parts. In the
case of the blood corpuscle the action takes place as follows :—
The original bell is large, but of a soft, flaccid skin-like
texture ; but it is not able to take on any particular form while
it is in the crush of marrow cells, but when it gets into the
stream of blood, its peculiar elasticity shows itself in such a
way that it endeavours to attain the bell shape inherent in it
from its origin. Then small, round, dark red blood corpuscles
are formed, which are gradually rolled out to form the larger,
flat biconcave discs.''*

The views here propounded as to the cause of biconcavity
are, I think, too mechanical, and very unlikely to command
general acceptance. It is stated that so long as these bodies
are in the bone marrow, and are being pressed upon by surrounding structures, they cannot assume their normal shape ;
but that when by passing into the blood-stream they become
relieved from external restraint, they assume by means of their
peculiar *elasticity* the bell-shape inherent in them, and notwithstanding the presence of this elasticity, it is argued that the
rollings and impactions to which they are subjected convert
them in the first place into globules, and in the second into
biconcave discs. What becomes of their natural elasticity
while this moulding and modelling goes on we are not told,
and yet it is obvious that only plastic substances, devoid of
elasticity, like putty or clay, could be at all permanently

---

* These considerations flow out of the theory put forth by Rindfleisch that the red blood corpuscles are formed from the exterior matter
of red nucleated cells, like those found so plentifully in the blood of
the early embryo. For my views on this point I must refer the reader
to Section VIII. of this work, " On the Bone-Marrow."

influenced by such processes. The power of taking on these biconcave forms really rests upon a physical law which is competent to act under favourable conditions, with much greater precision than any such irregularly operating mechanical forces could possibly do.

Hunter and many of his contemporaries believed that the red corpuscles were *liquid bodies*, and there seemed nothing against this view so long as these bodies were considered to be *spheres ;* but the case was altered when it came to be found that their normal form was not spherical, but discoid.

After demonstrating in various ways their flat shape, Hewson expresses himself as follows : " These experiments not only prove that the particles of the blood are flat and not globular, but likewise by proving that they are flat they show that they are not fluid, as they are *commonly believed to be*, but on the contrary they are solid, because every fluid swimming in another, which is in larger quantity, if it be not soluble in that other fluid, becomes globular. This is the case when a small quantity of oil is mixed with a larger quantity of water, or if a small quantity of water be mixed with a large one of oil; then the water appears globular, and as these particles are not globular but flat, *they must be solid.*"*

This reasoning was deemed conclusive, and the difficulty all along experienced has been, how to explain the flowing liquid-like properties of the corpuscle, and at the same time account for its retention of a peculiar form unlike that in which minute liquid masses are usually seen, and for the remarkable elasticity by which it returns to that form after disturbance. This was formerly sought to be explained by assuming the existence of a delicate, more or less restraining capsule, and more latterly by the hypothesis of a colourless, structureless, transparent, filmy frame-work or stroma, which is supposed to become infiltrated in all parts with hæmoglobin.

This idea of a skeleton frame-work appears to have arisen with Rollett, who conceives that a stroma or matrix enters into

---

* Essay on " Red Particles." Page 220, Hewson's Works.

the structure of the coloured elastic extensible substance of the
red blood corpuscles, and that to this the form and the peculiar
physical properties of the corpuscles are due. It is supposed
that the colouring matter can be separated from the stroma
without causing the latter to lose its essential characters.*

. Thudicum defines this *stroma* as merely a name for a
substance which is supposed to give the blood corpuscles *shape*,
and which remains when the other bodies are extracted, a kind
of *chemical skeleton*, which seems to be very different from the
albuminous matters combined in hæmoglobin, for it is soluble
in ether, alcohol, and chloroform, when these agents are
dissolved in serum.†

The blood corpuscles possess in a very marked degree
extensibility and retractibility, properties which, when present
together, constitute elasticity, and which are most perfectly
displayed by liquids having a certain degree of viscosity, *e.g.*,
soap-films, etc.‡ If the form of the corpuscle is due to a
stroma, then these properties must necessarily reside in the
stroma; but as they are the properties of liquids as opposed to
solids, we may fairly ask how is the possession of such
properties by a stroma to be reconciled with its assumed power
to coerce and maintain in a biconcave shape, a body, which we
know can take on the form of a sphere, with so slight a modification of its surrounding as that produced by the mere act of
shedding the blood.

If the stroma were rigid enough to maintain the biconcave
form of the corpuscle, in opposition to the sphere-forming
energy of cohesion, its liquid-flowing character would certainly
be restrained by the presence of such a framework.

The extensibility and retractibility of blood corpuscles is, as
before said, that of liquids, and the elasticity of such a substance

---

* *Vide* Page 408, Vol I., "Sticker's Human and Comparative
Histology." Article on the Blood Corpuscles, by Rollett.

† *Vide* Page 29, "Chemical Physiology."

‡ *Vide* the Author's Inaugural Address to Birmingham Philosophical Society, 1876. "New Researches on Elasticity and Contractility."

as caoutchouc is by comparison rigidity itself. If we could separate this substance into minute masses of the size of red corpuscles, they would, when submerged in liquid, appear as absolutely rigid bodies. The small motor influences which are in operation in a specimen of blood, such as currents, impacts, etc., would not affect their form in the least. Even oil globules of the same size as the blood corpuscles are inextensible under such delicate influences, and if the so-called stroma was not more limpid than oil the corpuscles could not possibly display any elasticity.

As before said, the elasticity of the corpuscles is peculiar to liquids, and even to comparatively limpid liquids, and therefore this so-called stroma must be liquid—more liquid than oil—and hence could do nothing towards maintaining a specific biconcave form in opposition to cohesion.

The fact that the hæmoglobin may be dissolved out by aqueous liquids and a framework remain which requires for its solution the usual solvents of fats or waxes, is nothing whatever to the point. In what respect does this differ from the fact that a piece of annealed steel of any form may have the most elaborate designs engraved upon it, and then all the metallic iron be absolutely dissolved away, leaving both form and design perfectly intact, but now no longer in steel, but in the carbon which was associated molecule for molecule with the iron ? * Have we not, too, the familiar example of bone from which either the earthy or the gelatinous matter may be removed at will without destruction of form ? Shall we speak of the carbon stroma of the steel, and say that it was previously infiltrated with iron ? Clearly all that we have any right to infer from these chemical differentiations is, that substances soluble in water, and others soluble in ether, alcohol, etc., are in the corpuscle in the closest possible combination or association, molecule for molecule, and that the one may be dissolved out, leaving the other *in situ* to maintain the integrity of the form.

* *Vide* the Author's paper on "Certain Molecular Changes in Iron and Steel during the separate acts of Heating and Cooling."—Proc. of Royal Society, London, 1877.

The fact is that neither the view of the existence of a
capsular membrane or of a stroma is at all necessary to the
explanation of the biconcave form, and there does not appear to
be any sound reason why we should not return to the view of
Hunter and his contemporaries, and regard the mammal corpuscles as *liquid bodies.* This we can assuredly do if we can
show that the peculiar biconcave form they usually possess is
consistent with liquidity.

The idea that one liquid submerged in another, with which
it is immiscible, must necessarily take on the globular form,
has had its origin in the circumstance that our only illustrations of this phenomenon have been derived from the observation of mixtures of oleaginous and aqueous liquids, and these
invariably give rise to spherical forms. There are, however,
certain general properties of organic substances in relation to
each other which have hitherto been overlooked, and which
will probably be found to have much to do with phenomena of
this kind, and with the chemical and physical basis of organic
development and growth.

Both aqueous liquid globules and biconcave discs can be
formed by a process which we may designate *liquid precipitation.*
There are certain combinations of organic substances which,
when in solution in water, possess such antagonistic relations to
each other that they refuse to remain together in the same
solution, and as a consequence the one for which the liquid has
the greatest affinity will displace the other ; nevertheless the
displaced substance will retain to itself a portion of the water
in which it was originally dissolved, and as a consequence of
this, it does not come down in a powdery amorphous form, but
in the shape of liquid colloid globules or discs, as the case may
be, so that we get aqueous solutions of organic substances set
free in an independent form in other such solutions, the one
being immiscible with the other, just as in the case of oil and
water. The one organic liquid bears, in fact, a formed or morphological relation to the other, and this relation may take on
the character of corpuscle, fibril, or film. Which of these will

occur in any particular case, will depend upon the degree of liquidity, or viscosity of the solution, which separates itself, or is precipitated. The corpuscular state is associated with the greatest degree of liquidity, while the film and fibre are the products of a more viscous condition. If, for example, pieces of sheet gelatin be immersed in a cold ammoniacal solution of carmine, a certain portion of carmine unites chemically with the gelatin, as is proved by the fact that the reddened gelatin may receive subsequently a protracted washing and soaking in water without imparting to the latter the least colour. By modifying the strength of the carmine solution or the amount of gelatin we may obtain gelatin containing a larger or smaller amount of carmine in combination. We will call this preparation carmino-gelatin. By the aid of heat it may be dissolved in water, and if, after cooling, a little finely powdered gum acacia be dissolved by shaking in the same solution the carmino-gelatin comes down in the shape of minute liquid globules of tolerably equable size, which have a fair resemblance to spherical blood corpuscles, and float in a colourless liquid. It was also found that when gelatin was added to an alkaline solution of hæmatin it combined with it in the same manner as the carmine, and yielded a liquid globular precipitate on the addition of gum acacia. A similar result was found to take place with globulin. Portions of the crystalline lens were soaked in a solution of hæmatin; and this solution being thickened with gum acacia, transparent globules resulted as in the previous cases. The size of the corpuscles could be varied by the strength of the solution, the weakest solution producing the smallest corpuscles. In the previous cases the precipitated forms were globular; but if powdered gum acacia be added to a solution of Nelson's *Opaque* gelatin, the opacity of which appears to be due to a finely-divided material allied to myelin, biconcave corpuscles and biconcave granules are precipitated, such as are seen in Photograph 40, Plate VIII. These artificial discs appear to be formed of myelin, a substance which combined with globulin enters into the constitution of both lymph and red-blood corpuscles. By treating lymph corpuscles with strong acetic acid the myelin may be made to separate and show itself in refractive *annular* masses. (*Vide* Photograph 41, Plate VIII.)

The remarkable properties displayed by myelin at once relieve us from the necessity of considering that one liquid or solution submerged in another must inevitably take on the globular or spheroidal state. The fact is, this substance appears to represent the extending or spreading-out tendency, as opposed to the gathering-up or sphere-forming property. The biconcave and the annular forms seem to be related to a kind of balancing of these two properties. These forms do not depend upon one condition alone, but on several, some of which rest in the body itself, and others in the surrounding medium. This is shown by the circumstance that the biconcave form is capable of either being obliterated or intensified by alterations in the environment. I do not wish for one moment to imply that blood or lymph corpuscles are originally formed in this manner; on the contrary, I know them to be directly descended from cellular structures; but I consider that aqueous relations such as those existing here must also be present in their case, and, in fact, in the case of young structures generally. Were it not so I can conceive of nothing but their immediate solution and disappearance. It is by some such relations as these that bodies so liquid-like in their character are able to preserve their integrity in aqueous solutions, and this view is supported by the fact that it is only in liquids specially related to them that the blood corpuscles can maintain their existence; for it is well known that the corpuscles of one animal undergo solution in the serum of another. I consider, too, that the form which these bodies assume is as dependent upon the constitution of the liquid in which they are submerged as upon their own, and it is on this account that we so frequently see numbers of the mammal blood discs become spherical after the blood is shed, a change which, it may be remarked, is largely confined to the younger corpuscles. It is also a fact that has not received sufficient attention, that alterations in the serum will, on the one hand, change the red corpuscles into large, pancake-like discs, in which their concavities are nearly obliterated, and on the other will so intensify the biconcaving tendency as to convert them into perfect rings. It would, therefore, seem that the biconcave form is to be regarded as *an arrested annular form*. This annulating property belongs to the corpuscle as a substance, for it occurs in fused masses of corpuscles, in single corpuscles, and in fractional parts of them.

Under the conditions in which blood is usually examined very little is seen of this annulating property, because it is the younger, colourless, and slightly coloured discs which have the greatest tendency to display it, and these usually sink down into liquid films upon the surface of the slide; but if we take measures to prevent, or to partially prevent this melting down action, we have formed little *liquid pools*, which result from the fusion or coalescence of these younger discs. Such a condition is exhibited in Photograph 21, Plate V., and is obtained by running blood into osmic acid, already occupying the space between the cover and the slide. It will be noticed that the tendency of corpuscles to fuse and spread is inversely as their colour. If, after these little pools of corpuscular matter have been rendered sufficiently viscous by the osmic acid they are treated with a little water, they immediately absorb it, and commence to expand, and at the same time the extreme edge becomes raised, rounded, and thickened. The more coloured corpuscles which at first simply lay in the fused liquid-like mass have now become fixed in it, and as the substance expands they are removed to a greater distance from each other. Photograph 137, Plate XVIII., illustrates these conditions. This specimen has been subsequently stained with an alcoholic solution of aniline brown, in order to render the delicate *colourless masses* more apparent. As this annulating action proceeds the material is gathered more and more to the periphery, till at length a ring alone remains, as seen in Photograph 138, Plate XVIII., and this, then, sometimes splits at its weakest point, and irregular fibres result, such as are shown in

Photograph 142, Plate XVIII. Single corpuscles also can in this way pass *beyond* the biconcave form and become simple rings, and ultimately curved fibres. The circumstances which contribute to this action appear to be such as tend to render the corpuscles a little more viscous, as when these changes occur they are seen to have an increased tendency to adhere to each other, and as a consequence an appearance of chain mail is sometimes produced. (*Vide* Photograph 139, Plate XVIII.) In this case very weak tannic acid was the agent employed to reduce the limpidity of the corpuscles. In this example it is the older or full red discs which are affected in this manner. In Photograph 140, Plate XVIII., young, colourless discs which have been rendered less limpid by osmic acid may be seen undergoing this action. It will be noted that very perfect ring forms are produced, while the more coloured corpuscles, as often happens from the use of osmic acid, have undergone darkening and contraction.

These ring-forms have been withdrawn from the blood by a method of isolation.* As with the masses, so with the single corpuscles : when the annulation proceeds beyond a certain extent, rupture may occur at some point of the periphery, and the corpuscle becomes now converted into a more or less curved fibre. Single rings of this character, which have been rendered plainer by staining, are seen in Photograph 138, Plate XVIII.

When corpuscles lying side by side become fused together it not unfrequently happens that *holes* of considerable size form in them. This appears to be due to the perforation and gradual dilatation of their natural biconcavities or depressions. (*Vide* Photograph 141, Plate XVIII.)

Now these effects which we have observed in blood corpuscles, and which are obviously due to the extreme action of the principles upon which their biconcavity depends, are by no means peculiar to them, but may be seen operating in connection with other substances as the mere result of physical

---

* Curved cover glasses are used which can be gently tilted from side to side, the younger corpuscles are found attached to the cover.

law. If, for example, we take minute portions of the solid fats, such as butter or lard, and place them upon the surface of water of a temperature sufficient to just melt them, but not to render them too limpid, we shall notice that they first form discs, the size of which will depend upon the amount of the material taken. If a small disc only is formed it will be seen that after a time a slight depression will occur upon its surface, which in a brief space will become a minute hole ; this hole will gradually enlarge till a perfect ring is formed ; this ring will, in the event of one point being weaker than the rest, break and extend itself into a line, which will then be drawn up again into a disc, as at the start. This process may often be seen to repeat itself several successive times ; indeed there is reason to think that it would continue to do so indefinitely if the surface of the liquid remained pure and clear. Upon the other hand, if the disc or film of grease be large, depressions and, subsequently holes will appear all over its surface, and it will become a reticulum, as shown in Photographs 143 and 145, Plate XVIII. The splitting of the rings may be seen on the upper edge of Specimen 143, which was solidified by cold before the separated parts could be retracted.

These are capillary effects, and are due to tension surfaces, and would not occur in the case of the fat if it were submerged and unconnected with surfaces, because in the fat the sphere-forming tendency is greatly in excess ; but if we take a material such as myelin, in which the tendency to form spheres has been nearly overcome, and the disposition is rather to extend or spread, we have no difficulty in obtaining similar results under conditions of complete submergence, and in producing biconcave discs and ring forms in abundance.

Photographs 144, 146, and 147, represent biconcave and ring forms obtained by spreading myelin into a thin film upon a glass slide, and, after placing over it a cover-glass, running in saturated solution of sodium chloride. These bodies may advantageously be compared with the various conditions of the blood corpuscles photographed in the same plate. The large myelin rings, seen in Photographs 144 and 146, compare

well with the blood-rings seen in 138, 139, and 140. The
mode of union or adhesion displayed by these myelin and blood-
rings are strikingly alike, as is seen in Photographs 138, 139, and
146. The myelin forms in Photograph 147, also compare well
with the modified forms of the blood corpuscles in Photograph 148. In Photograph 146 the smaller myelin forms
compare well with the biconcave discs of the blood, and support
the previously obtained results in Photograph 40, Plate VIII.
The series of changes which we recognise consist in the
passage of a globule into a disc, of the disc into a biconcave
form, of the biconcave form into a ring, of the ring into a fibre
or rod, and of the fibre or rod into a globule again, and this
action will repeat itself any number of times, providing
the surface of the liquid keeps clean or the ring does not happen
to be formed of equal strength at every point, and thus
balance the annulating force. In that case the action
will not proceed beyond the ring, so that given a *viscous
liquid disc* there are two opposite ways towards which it
may tend to regain equilibrium ; but in doing this it may remain
in and exhibit the forms which are intermediate, and this is
controlled or determined by the nature of the surrounding. If
its tendency be directly towards the sphere it may exhibit
the various stages of convexity ; if towards the sphere
through the ring-form the various degrees of biconcavity.
The surrounding medium is the power which determines
these opposite tendencies to change of form, either in the
direction of the annular or the spherical form ; for if the surrounding medium is altered the corpuscle is handed over to one
or other of these tendencies. *The form, therefore, depends upon
a delicately balanced relation of the physical substance of the corpuscle and of its surroundings.* The substance of the corpuscle
is itself in such a state that it has little tendency on its own
account to take on either one form or the other. This condition
is obtained by counteracting the retractile energy which
is most powerful in *limpid* liquids by rendering them
viscous, which means easily extensible, the extensibility

of a liquid being opposed by its retractile energy.
That the retractile force of a liquid substance may be
almost entirely balanced or counteracted may be seen in
such a substance as myelin, which when submerged in water
exhibits the most marked tendency to extension under the
slightest perturbing influences, and this quite apart from any
relation to surfaces. On the other hand, the following experiment will show the power which a different external medium
may exercise over the retractile force of a moderately viscous
liquid substance, which is submerged in it : Produce upon an
iron ring, say two inches in diameter, a film, by dipping it into
a solution of albumen ; this, of course, will be a tension film,
and would, like a soap film, burst with any interference with its
continuity. Carefully submerge this film in turpentine, and it
will instantly lose all its tension, hang down like a bag, and
undergo great distension, with the slightest up and down movement of the ring in the liquid ; being taken out of the
turpentine it will immediately resume its state of tension, and
no amount of up and down movement in the air will distend it.
If pricked it bursts into mere vapour like a bubble ; on the other
hand if its continuity is interfered with while it is submerged, and
it becomes separated from the ring, it retains its film-like character,
and sinks to the bottom of the vessel. Here, then, we have an
example of the conversion of a fugitive tension film into a
homogeneous permanent film or membrane entirely by virtue of
its new surroundings, and only so long as these special surroundings are maintained. In view of such a fact as this, may
we not say that a structure is the same structure only so long
as it remains in the same surroundings ? If now we revert to
the advanced lymph corpuscle we find that it leaves the blood-
glands as a comparatively pasty and rigid disc, and in the more
liquid regions of the larger lymphatics and the blood vessels
gradually takes on a comparatively limpid character, and obtains
that degree of retractile power which, while it is consistent
with the assumption of the biconcave form, is essential to
its elasticity. As the lymph corpuscle is a *nuclear disc* when it

begins its independent career it has simply to pass, in obedience
to the principles explained, from the disc to the biconcave
form.

Whatever may be said as to the correctness of the explana-
tions here given, one thing is absolutely certain, viz., that viscous
liquids exist, which, when submerged in other suitable liquids,
give rise to the formation of delicate biconcave discs, which are
altogether unconnected with surfaces, and float freely about,
and exhibit elasticity and the same general physical properties
as the young blood corpuscles.   It will, I think, be found that
these phenomena are confined to microscopic masses of matter,
i.e., to matter of a peculiar constitution in a fine state of
division, and which is thus relieved in great measure from the
influence of grosser forces, e.g., gravitation.*

The view that the blood disc is a liquid body, and that its
biconcave shape is due to the operation of physical conditions
and not to structural restraint, does not in any way preclude
the possibility of the existence of a slight difference between its
exterior and interior ; indeed, it is most likely that the exterior
will be modified by constant contact with other substances in
the liquor sanguinis, and that a species of pellicle will result.
Nevertheless, all that can be claimed in relation to these two
constituents is, that the exterior is a little more solid or pasty
than the interior; not that it is to be regard as a true double
contour membrane or capsule, but rather as the surface matter
modified for some slight distance inwards by contact with the
liquid which surrounds it; that, in fact, the whole structure was
originally a smooth colloid particle in a saline colloid liquid ;
and that the latter has reduced its external surface to a thicker
or less limpid state, forming, so to speak, a *physical membrane*,
liable to changes and alternations from the less liquid to the
more liquid state, and *vice versa*.   In fact both the blood and
the lymph corpuscles are to be regarded as minute particles of
protoplasm which have undergone differentiation by the formation

* Everyone knows how slowly matter in a very fine state of division,
and indisposed to aggregation, sinks even through a limpid liquid.

of an exquisitely delicate physical pellicle. In such a state of things the transition from solid to liquid would be very graduated, and whilst liquid could pass from and to the interior it would constitute no evidence of an essential difference between it and the exterior. Such a difference of constitution would permit osmotic phenomena to come into play, and thus the various *irregular* modifications of form of which the blood corpuscles are susceptible when the serum is modified would be readily explained.

Part V.

On Fibrin Formation and Coagulation.

I have stated that the discovery of *colourless discs* in the blood not only enables us to explain its mode of development, but also the manner in which fibrin is formed. Hitherto little attention has been paid to the microscopic aspect of this question, mainly because the formation of fibrin has been considered rather as a chemical than as a physical and morphological problem. As neither the red nor the white corpuscles could be regarded as the fibrin stroma, the only tenable supposition was that the element or elements which yielded fibrin must be *in solution* in the liquid of the blood. During the course of my microscopical researches, I have accumulated a considerable amount of evidence which goes to show that *the fugitive discs of the blood* produce by their degenerations and transformations the delicate networks, films, and masses which constitute the fibrinous deposits of the blood. In Photographs 21 and 22, Plate V., evidence is adduced of the manner in which the *colourless discs* fuse together and form pools of clear, transparent *liquid*, which is immiscible in the liquor sanguinis, and which may either sink down and become hardened into fibrin films, or, if agitation be present, be drawn out into fine filaments, which by their interlacements form fibrinous networks, and these subsequently become hardened. Photograph 60, Plate XI., illustrates the direct conversion of colourless discs into delicate fibres by the combined action of annulation and extension by currents. In Section IV., in which the views of M. Hayem are discussed, numerous examples are given of the modes in which fibrin is formed from *the corpuscles of the fugitive group.* It was my intention to discuss here the microscopic aspects of coagulation, but the subject is one of such wide extent, and its microscopical, chemical, and physical bearings are so inseparable as to render a separate work necessary. I shall, therefore, content myself by stating, in opposition to the existing views, that fibrin has a morphological basis in the blood, and is always present in the shape of *its colourless and slightly coloured discs,* that these are, in fact, the true fibrin stroma. When blood is kept fluid by the reduction of its temperature to 0° Cent. the red corpuscles subside, but the colourless and slightly coloured discs, and the ordinary white corpuscles, remain in suspension, and the former are mostly in a disintegrated and granular condition. If solid neutral salt be added to such plasma, the granular matter which is precipitated, and to which the name of *plasmine* has been given, is nothing more than colourless and slightly coloured discs which have undergone inspissation and granulation, and having thus acquired a greater specific gravity than they ordinarily possess, sink to the bottom of the vessel. There is no precipitation in the sense of the passage of a substance from a soluble into an insoluble state, and, on the other hand, when this precipitate is diffused into weak saline, it does not, as is usually supposed, enter into solution, but simply becomes transparent and hyaline by the absorption of the liquid, very much after the fashion of the insoluble portions of gum tragacanth, and in this condition it is (like the colourless discs from which it was originally derived) invisible, because it possesses the same refractive index as the saline solution with which it is incorporated. Subsequently it becomes hardened, and the liquid, viscous state gives place to a gelatinous one. For the discussion of the nature, mutual relations, and behaviour of the substances known as Fibrinogen, Fibrinoplastin, and Ferment, I must refer the reader to my full paper.

Further Researches on the Third Corpuscular Element of Mammalian Blood.

In my first communication on the development of the blood, I drew attention to the fact that there existed in the blood of mammalia a very large number of *colourless discs* which had escaped recognition, and stated that an intimate knowledge of the origin, nature, and behaviour of these bodies was absolutely essential to the comprehension of the physiology and pathology of the blood, and also of the phenomena connected with its coagulation. These conclusions were not hastily arrived at, and as continued and unremitted study of the subject has simply served to strengthen them, I feel it to be my duty, in view of the important issues at stake, and of the fundamental changes in opinion which the discovery involves, to patiently and carefully weigh any objections which are, from time to time, submitted to me by competent persons. I propose, therefore, to consider here the various objections which have come to my knowledge, and, at the same time, to endeavour to justify the title of the paper by describing a number of new experiments.

It is gratifying to be able to state at the outset that my task has been much simplified by the fact that, so far, everyone who has followed the methods which I have employed has succeeded in obtaining the results and appearances described and illustrated by the photographs.

Such observers are, therefore, prepared to admit "the existence of *colourless transparent discs,*" and the issue is consequently narrowed to the points, as to whether such discs exist in normal unshed blood, or whether they come into existence during the mere act of shedding the blood, or are in any way produced by the methods used to display them. The discussion, therefore, hinges upon the question as to what these *colourless discs* really are, and, fortunately, it is only open to us to accept one of two views—they are either colourless corpuscles which are destined to become coloured, or they are red ones which in some way have lost or been deprived of their colour.

In support of the first view, it can be urged that we already know that large numbers of bodies of the same size as these discs continually pass into the blood from the lymphatics. I have shown that these bodies are not globules but discs, that their plane surfaces are covered with depressions, and that they are even capable of becoming biconcave, in some instances, while yet in the lymphatic glands.

Although these bodies are known to pass into the blood in such numbers, their presence has never been recognised in this fluid, nor, to my knowledge, has anyone, with the exception of Professors Gulliver and Bennett, in England, and Kölliker, in Germany, troubled about their absence.*

---

* Professor Gulliver, who has taken much interest in the progress of this research, and from time to time favoured me with suggestions and criticisms, has recently drawn my attention to the apparent ambiguity of this sentence.

He says, "With Howson, it was a cardinal tenet that the lymph corpuscles enter the blood, and could be seen there, and I never knew that any physiologist doubted the fact, till I looked at page 38 of your paper."

The confusion here seen arises entirely from the practice of having a common designation for the white blood corpuscle and for the true lymph corpuscle.

It seems to me most desirable that we should retain the name white blood corpuscle, or leucocyte, for those elements of the blood

A *colourless disc* is now found in the blood, which I shall be able to show has the properties of a lymph

---

glands which are seen without difficulty *in the blood*, confining the term lymph corpuscle to those products of the blood glands which undergo direct conversion, first into colourless discs and subsequently into coloured discs or red corpuscles. These bodies do not appear at all, as far as the blood is concerned, till they show themselves as pale coloured blood discs.

Professor Gulliver himself distinguished most carefully between these two kinds of lymph products (not in the blood, where they do not appear in contrast, but in the lymph current and in the blood glands,) when he says, " The globules of the chyle, of the thymus fluid, and of lymph are smaller and differ in structure from the pale globules of the blood. In these last there are two, three, or four nuclei, easily seen when the envelope is made more or less transparent or invisible by acetic, sulphurous, citric, or tartaric acid. But the globules of chyle, of lymph, and of the thymus fluid, like the nuclei of the red corpuscles of the oviparous blood, are only rendered more distinct and slightly smaller by any of these acids, so that the central part presents no regular nuclei or divided nucleus such as are contained in the pale globules of the blood. In short, these last-named globules have the characters of perfect elementary cells, while the former globules, as shown in the note to ' Gerber's Anatomy,' page 83, resemble, and probably are, nuclei or immature cells."

He gives, too, the measurements of the pale globules of the blood and of the true lymph corpuscles as 3,000th as against 4,600th of an inch ; *vide* pages 243 and 244 Hewson's Works. It was the pale globule, or white corpuscle of the blood, which Hewson saw and referred to in the following historical passage, " Secondly, we have proved that vast numbers of (central?) particles made by the thymus and lymphatic glands are poured into the blood through the thoracic duct, and if we examine the blood attentively we see them floating in it." Page 282 section 98, Hewson's Works. To this Professor Gulliver appends the following pertinent note :—" This passage is so clear as completely to set aside the claim made of late years, by M. Mandl and others, to the discovery of the pale globules of the blood. In that of mammalia, it is quite evident, that Hewson had seen these globules, and considered them in all the vertebrata as lymph corpuscles, a view which has recently been revived. Senac also appears to have seen the pale globules in the blood, and to have regarded them as belonging to the chyle."

Again he suggests—" That the lymph-globule is an immature cell, which may change in the blood, and even in the thoracic duct or lymphatic vessels, into the larger and more perfect pale cell of the blood, is very probable."

Further, on page 254 of the same work, Professor Gulliver distinguishes between the white globule and the lymph corpuscle, and says, " In the corpuscle, and not those of a decolourised red disc, and between this and the fully-matured red corpuscle, others having every intermediate shade of yellow tint, can be observed.

My view, therefore, is that these *colourless discs* are the lost *lymph or gland corpuscles,** which undergo gradual conversion into red corpuscles, by the secretion within them of colouring matter, after the analogy of the ovipara. As all this goes on in a region of invisibility, it accounts for the exceeding difficulty which has attached to the comprehension of the development of the blood, and removes that " blot " upon our scientific manhood which Rindfleisch has recently so much deplored.† As before stated, the existence of these corpuscles once admitted, the only other interpretation which can possibly

---

blood, besides the common pale cells, there are a few smaller corpuscles, like those of lymph : while in the larger lymphatics and thoracic duct there are corpuscles identical in size and structure with the common pale globules of the blood." The facts are these : In the blood, white corpuscles are readily seen, which possess several nuclei, and are proportionately large ; other smaller ones, in scantier numbers, which are uni-nuclear, but in which the nucleus is not readily brought into view by acids, because it fills up the entire cell—it is easily, however, displayed by imbibition and staining. This uni-nuclear cell corresponds to the true lymph corpuscle, in the first stage of its development. The multi-nuclear blood cell is developed from this body, in accordance with the ordinary processes of cell growth. This is the minor *rôle* played by the true lymph corpuscle. By far the greater majority of these uni-nuclear lymph cells become in the glands dispossessed of their cell walls, and elaborated into smooth, transparent disc-shaped nuclei, which, becoming biconcave, enter the blood as its invisible colourless discs. The white corpuscles, multi-nuclear and uni-nuclear, therefore, as seen in the blood, represent the leukhæmic accidents of . the true lymph corpuscle, and when this divarication exceeds a certain numerical limit, it constitutes leukhæmia, which is a pathological state resulting from the mere intensification of a physiological condition always present. For further information on this head, see the " Sections on the Products of the Blood-glands, and on Leukhæmia."

* In this term the corpuscles of the spleen are included.

† Ueber Knöchenmark und Blutbildung von Dr. G. E. Rindfleisch, Professor in Würzburg.

be put upon their presence is that they are discs which have yielded up their hæmoglobin. Three conceptions are here possible :—1. They may have become decolourised during the mere act of shedding the blood. 2. They may represent a stage in the dissolution of the red disc. 3. They may be produced by the methods employed to reveal them.

What I have to say respecting the first and second conceptions will be found on page 17 of my previous paper. I shall here confine myself to the consideration of criticisms which have been made against my methods. It must be remembered that, in order to sustain the position that these colourless and intermediate coloured discs are simply decolourized red ones, it is necessary to consider all my methods, *however various*, as leading to one and the same result, viz., to a complete discharge of the whole of the colouring matter of some of the corpuscles, and a partial and graduated discharge from others, so as to present an entire and perfect series ranging from a colourless to a full red disc. According to my view, the biconcave corpuscles do not become visible in the blood till they have acquired just that slight amount of colour which is necessary to enable them to be distinguished from the Liquor Sanguinis. Below such barely visible corpuscles there are, therefore, numbers wholly invisible, and the palest of the visible order may, under the influence of certain causes, lose their colour, and join the invisible group. These corpuscles I have frequently observed to disappear. It is not with these, however, that we have to deal, but with corpuscles that are never visible under any of the usual methods by which the blood is examined, and which require the most delicate precautions to be adopted for their physical preservation.

This defence of my methods has been rendered desirable by an abstract and criticism which appeared on January 15th, 1880, in the "London Medical Record."* The abstract itself

* A. Hart.

is fairly accurate, but the criticism, although not wanting in ability, shows that the author has failed to thoroughly assimilate my views, for it is based entirely upon misconception both of the nature of the methods and of the actual procedures adopted.

In common with others who have followed the directions laid down, the author of this report has had " no difficulty in obtaining identical appearances with those which he (Dr. Norris) has described and photographed." But it is stated " that those appearances are to a considerable extent due to artificial causes, and that Dr. Norris's methods appear in a great number of instances to be open to the objection of tending to produce the appearances which he describes as normally existing." It is further affirmed that " Dr. Norris adopts *three* methods by which the colourless corpuscle is rendered visible." 1. By changing the refractive index of the serum by the addition of a saturated solution of common salt. 2. By withdrawing the serum from the corpuscles by the method which he calls " packing." 3. By withdrawing the corpuscles from the serum by the method which he describes as that of "isolation." Thus at one stroke the number of my methods is reduced from *ten* to *three.**

The next step of my critic is to show " that saturated solution of salt is capable of dissolving some of the hæmoglobin from the red corpuscles, thus rendering some of these faint and ghost-like." It is admitted " that this fact is well known to histologists, and is mentioned by Rollett in Stricker's manual." This is followed by the remarkable statement :—" It is to be noted that Dr. Norris recommends that, *in all cases in which the blood is to be examined for these transparent corpuscles,* it should be passed immediately it is drawn into an ammoniated

---

* In my previous paper I gave ten distinct methods by which the existence of these corpuscles could be demonstrated :—By impact; by packing ; by isolation ; by osmic acid vapour and cold ; by solution of osmic acid ; by alcohol ; by altering the refractive index; by spreading and drying ; by staining ; by tinting the Liquor Sanguinis.

solution of carmine in a saturated solution of common salt, with the object, as he states, at one and the same time of preserving and staining the colourless corpuscles. *This process he recommends as preliminary to examining the blood by the methods of 'packing' and 'isolation.'* It would be interesting to learn from Dr. Norris if all the photographs in his paper were taken from specimens of blood which had first been subjected to the action of saturated solution of salt."

The gravamen of all this is, saturated solution of salt will produce colourless discs. Dr. Norris used this in *all his processes*, therefore their presence is easily understood.

The fact really is that in *two* only of the *ten* methods has salt been employed, and in *three* only out of forty-eight photographs was it used. I have most sedulously avoided the use of re-agents of every kind *as a means of demonstrating the existence of these corpuscles*, knowing full well the doubt and difficulty which would be introduced into the question by so doing. I distinctly stated this in my paper in the following terms :—"The utmost caution is of course demanded in drawing conclusions after the addition of re-agents of any kind to the blood—such a method is always open to the objection that the re-agent may produce that difference which is simply sought to be displayed." The existence of the new corpuscles having been first demonstrated by unobjectionable methods, it was then, of course, permissible to see what could be done by staining, and by alteration of the refractive index of the liquid surrounding them—in short what could be brought out by re-agents generally.

In the methods of " impact," of " packing," of "isolation," and of " spreading and drying by air," *no re-agent whatever is added to the blood*, and I am convinced that as methods they are quite unassailable.

The uncertain state of mind of my critic is rendered sufficiently obvious by the fact that four mutually destructive theories are propounded to explain the presence of this corpuscle.

1.—It is due to salt.

2.—To pressure or compression.

3.—To violence.

4.—To *post-mortem* changes.

The latter view is enunciated in the following terms :—" I may be permitted to say, after going with care through a series of observations on this subject, I am disposed to believe that the colourless corpuscles which (without the addition of a saturated solution of common salt) are still undoubtedly seen when the blood is examined by the method of ' isolation,' are red corpuscles which have undergone *post-mortem* changes prior to taking part in the formation of fibrin."

On this it may be remarked that if these corpuscles are red ones which have lost their colour after the blood is shed prior to being formed into fibrin, they must always be a constituent of *shed* blood, and it is therefore a work of supererogation to try to show that they are produced by the methods employed. My critic seems also to have forgotten that fibrin may form upon a foreign body, such as a thread passed through · a vein in a living animal. Any sound theory of fibrin formation must be competent to cover and explain every well-attested fact which experience has accumulated.

I shall now examine in detail these and other objections which have been made to some of my methods :—

Objection I.

Packing Method*.—On this method the writer remarks— " As to the effect of *pressure* on the corpuscles, when passed

---

* From a number of thin cover-glasses a slightly convex one is selected, by ascertaining that one surface (the convex) gives, by reflected light, a sharp image of the window bars, or gas-light, and the other a blurred and indistinct one. This cover is strapped, with its convex surface downwards, upon a microscopic slide, so firmly as to produce a series of Newton's rings of an elongated form. [Vide Plate I., Fig. 1 previous paper.] The two objects gained by this arrangement are as follows :—

between the glasses showing the rings of Newton—when the colourless corpuscle is first seen by this method, it is very striking. I closely observed what took place when the blood was treated in this way, and I finally arrived at the opinion that the *pressure excited by the glasses* is so great that it causes the corpuscles to discharge their hæmoglobin, and to become transparent* (colourless?)" The corpuscles are here subjected to an extreme degree of pressure. They are drawn by the force of capillary attraction between two glass surfaces, bound firmly together, until they reach a spot which they cannot pass, by reason of the close contact of the two glasses. They, therefore, become wedged in, and are subjected to the action of two forces—the capillary attraction, which is drawing the Liquor Sanguinis from around them, and the pressure above and below of the glasses between which they are tightly wedged. This pressure causes certain of the corpuscles to lose their hæmoglobin, and to become transparent. Some of the corpuscles are acted on in this way much more rapidly than others; but, in course of time, a great number of the corpuscles become quite colourless, and all of them, as they pass towards the centre of the glass, where the pressure is greatest, become paler. I have repeatedly watched this transformation take place." The writer is in error in considering that the red corpuscles are by this method subjected to strong pressure. The conditions are precisely those which obtain in

---

1. The glasses are everywhere in such close proximity as to admit the corpuscles flatways, and in single layer only.

2. In the part occupied by Newton's rings, the proximity is much less than 1-10,000th of an inch, and thus presents a *barrier* to the passage of the corpuscles, whilst allowing free passage to the Liquor Sanguinis. By this means the corpuscles are kept in the part marked A, whilst the Liquor Sanguinis is filtered off into B.

This packing of the red corpuscles causes them to mould themselves around the invisible bodies present.

* It is important in these considerations not to confuse the terms colourless, transparent, and invisible—they are by no means synonymous.

a capillary tube, with a diminishing diameter in one direction. It is more correct to regard this condition as one rather of coercion of form than of pressure, for the corpuscles are so plastic and flowing in their character, that their form undergoes modification with greater facility than is the case with many liquid globules, *e.g.*, globules of oil in water, and if they are compressed in one direction they extend with great ease in another. Of course, when very much flattened and thinned, they *appear* to have little colour; but if liberated from the coercion, they regain their form, and with it their original tint. It is easy to show:—

I. That the corpuscles are not by this method subjected to pressure in the proper sense of the word.

II. That coercion of the character to which they are sometimes subjected is customary to them in the capillaries, and is incompetent to remove from them their colour.

III. That this method will bring these colourless corpuscles into view when the proximity of the glasses is such that no *compression or coercion of any kind whatever is exerted.*

In the first place the strength of the current is the measure of the coercion, not the proximity of the glasses; or the latter only to the extent that it is related to the rapidity and strength of the current. The force of the current is measurable by its rate. The corpuscles will not proceed into interstices which would crush or injure them. In this respect the method differs from forcing glasses into greater and greater proximity. There is no pressure or compression of a kind which would lead to diminution of volume, as in squeezing a sponge in the hand. The force of the capillary current, and the elasticity or extensibility of the corpuscles, are the two opposing factors at work; when the current carries the corpuscle into a space limited in one direction only, and is incapable of carrying it further, it is because the force of the current is at this point balanced by the elasticity of the corpuscle, a stronger current would carry it further. The glasses being fixed and rigid do not exert any pressure upon the corpuscles, but simply afford the statical conditions, by means of which the current acts upon the corpuscle to modify its form, and when the current ceases, the proximity of the glasses prevents the elasticity of the corpuscle restoring the original form. Will .any one seriously affirm that the slight force exerted by the elasticity of the corpuscle, in its efforts or tendency to resume its normal form, is competent to cause it to discharge its colouring matter ? Corpuscles are constantly subjected to such influences and modifications of form, both within and without the vessels, and do not in consequence give up their colouring matter. I have often seen them flattened into pancakes or extended into fine lines without any such loss; in all cases, on the resumption of their original form, they assume their customary tint. I know of no method of decolourising a blood corpuscle, excepting that of solution of its hæmoglobin ; and even in this case many points arise, of which I shall speak further on.

Then again, supposing pressure to be present, acting equally upon the corpuscles in every direction, it could never discharge the colour from such bodies as the *mammal* blood corpuscles. This might occur if the colouring matter was simply entangled mechanically instead of being chemically combined in the substance of the corpuscle, or if we could regard the corpuscle as a *rigid* sac filled with colouring matter, we might imagine such a result possible ; but this is clearly not the case, for red corpuscles may be broken up into granules without loss of colour ; in fact, mechanical violence, *per se*, seems to have very little, if any, power to modify the colour of the corpuscles, and it would be surprising if it had, seeing that the colour is in chemical combination. But it is by no means necessary to subject the corpuscles to any pressure or compression whatever; it is simply required that the proximity of the glasses shall be such as to prevent the corpuscles rolling over or resting with their edges uppermost, or one corpuscle slipping above or beneath another—that is to say, the space between may be something more than the $\frac{1}{10,000}$th of an inch, but not twice this. In some of my best specimens the corpuscles have undergone no compression, as can readily be known from the fact that they have not increased in diameter beyond that of the normal biconcave disc. (Compare Photographs 1 and 4, Plate II.)

Again, local causes act locally, and if these colourless corpuscles were produced by compression due to the proximity of the glasses, they would occupy invariably the part nearest to the rings, but the fact is, fewer of these corpuscles can be seen near to the barrier, because they have formed *adhesions* to the glass before reaching this point.

They are distributed over the entire space, often right up to the edge of the specimen—in fact the peculiar adhesiveness of the corpuscle constantly causes it to be filtered out or held back before the rings are reached, and if the rings are sufficiently distant from the point of ingress of the blood, this will always be the case. The power of these colourless discs to adhere in this manner, (like the white corpuscle,) while the red ones remain unaltered and sail freely about and among them, is itself an evidence that they are something *different in kind*. Again, any point of arrest in any part of a specimen gives rise to packing, and displays these corpuscles—hence they are constantly seen in the little patches of packing which occur about air-bubbles, in parts of the specimen, where great freedom of movement *without modification of form* is possible. Side by side with these corpuscles are others which have obviously undergone no change of colour. *All are subjected to the same conditions.* Is it possible that some should be absolutely pressed white, while others in immediate contiguity undergo no change?

Photography is an excellent test for minute shades of yellow colour, and as it renders the extremes of these corpuscles, *black* and *white*, we know that the amount of hæmoglobin present in the first case is exceedingly great in comparison with that in the other.

The affirmation made by this writer of having repeatedly " watched the transformation of red corpuscles into colourless ones " while using this method demands more than a passing notice.

It is quite true that as the corpuscles pass within the outermost system of rings they become paler, and this because they become thinner, a fact which is known by the increase of their diameter.

As the Liquor Sanguinis becomes more charged with hæmoglobin (derived from the mass of corpuscles, and not from one more than another) the flattened corpuscles will become more and more obscured, indistinct, and faint, because the contrast between them and their surroundings becomes less and less. Under such circumstances they may become barely visible, and, perhaps, quite invisible, and this not because they have become colourless, but because the Liquor Sanguinis has become coloured up to a like intensity with themselves. For the determination of matters of such delicacy the average human eye is an inadequate colour instrument, and requires to be supplemented by photo-chemical tests. It is also impossible to observe these differences properly without a standard of colour.

By means of photography we can transmute with extreme precision these shades of yellow colour into degrees of black and white, which every eye can estimate with equal correctness.

The air bubbles present in specimens being absolutely free from colour, constitute an excellent standard of comparison for the inconstants, viz., the corpuscles and the Liquor Sanguinis.

The normal tint of the new corpuscles is closely that of the air bubble, and precisely that of the normal Liquor Sanguinis.

The normal Liquor Sanguinis is not quite as colourless as the spaces of the air bubbles, for as soon as blood is shed hæmoglobin begins to diffuse slowly into the Liquor Sanguinis, and, as time goes on, this liquid becomes deeply tinted, and if at the same time the red corpuscles become flattened by any means, they are lost to view in it.

The new corpuscle stains slightly with the Liquor Sanguinis, so that it is seen at its best when the tint of the plasma is nearest to that of the air bubble, *i.e.*, when it is freest from hæmoglobin.

These statements apply to blood under the usual conditions in which it is examined ; but it is quite possible to prevent the staining of the Liquor Sanguinis altogether.

I have ascertained that colloids, such as gum acacia and albumen, entirely prevent the exosmose of the hæmoglobin into the Liquor Sanguinis. They are therefore of the greatest use in testing for the presence of this corpuscle, and for the intermediates, for they *preserve all the colour in the corpuscles,* and therefore prevent the delicately-tinted intermediates from fading away into invisibility, by the combined effects of loss of colour on their own part and its gain on the part of the surrounding medium.

Colloid Method.—Reduce a little dry soluble albumen to[*] the state of fine powder, and having pricked the finger, stir up well with the blood, as it lies on the finger, a small portion of the powder ; run the blood beneath mica or glasses prepared for the packing method.

When the colouring matter is thus prevented from leaving the corpuscles, there is no fading away of the red corpuscles at the barrier ; or, to speak more properly, they do not become hidden in the stained Liquor Sanguinis ; and this is the case even after the specimens have been kept for eighteen hours.

This experiment, it must be borne in mind, is of a crucial character, for the coloured corpuscles do not yield up any of their hæmoglobin, yet the invisible corpuscle is present as usual. It cannot, therefore, be a decolourised red corpuscle.

Plate X., Photograph 49, is intended to show the contrast of colour between the air spaces and Liquor Sanguinis soon after the blood is shed. It will be observed that the Liquor Sanguinis is of a darker tint, owing to the presence in it of hæmoglobin, derived from the corpuscles.

---

[*] Soluble albumen may be obtained of R. W. Thomas, 10, Pall Mall, London.

Plate X., Photograph 50, is an example in which the colouring matter has been prevented from exuding from the corpuscles by the use of a colloid—in this case albumen. The air-spaces and the Liquor Sanguinis are here seen to be very nearly of the same tint, (*vide* upper left-hand corner in the intervals between the unpacked corpuscles.) The invisible and intermediate corpuscles are brought into view in the packed part of the specimen, and thus a demonstration is afforded that such corpuscles are not decolourised red ones, for no colour has passed into the serum from *any* of the corpuscles.

Plate X., Photograph 51, represents the appearance immediately contiguous to the rings of Newton. The Liquor Sanguinis, as compared with the air-spaces, is of a very dark tint, indicating the presence of much hæmoglobin. The red corpuscles have also increased in diameter, and have a fainter appearance. This faintness of the corpuscles is due to two causes ; firstly, they are flattened, owing to the proximity of the glasses ; and, secondly, the contrast between them and the Liquor Sanguinis is diminished, owing to the colourisation of the latter.

Plate X., Photograph 52, also represents a portion of a specimen near to the rings of Newton, but in this case the integrity of the red discs has been preserved by means of a colloid (albumen.) The Liquor Sanguinis is seen in this case to be very nearly as colourless as the air-spaces. The discs have lost no colour, and the serum has, therefore, gained none ; and, consequently, the corpuscles appear with their usual distinctness, and not as in Photograph 51. The great preservative power of the colloid is distinctly brought out by the fact that the specimen seen in Photograph 52 had been kept for eighteen hours, while that shown by Photograph 51 was taken within an hour of its preparation.

Plate X., Photographs 53 and 54, represent precisely the same spot, near to the barrier or rings of Newton, taken at an interval from each other of eighteen hours. In these cases no preservative was used. In Photograph 53, packed among the red corpuscles, may be seen here and there a few of the colourless discs ;

and it will be noticed that while nearly all the red corpuscles
have in Photograph 54 disappeared *in situ*, the colourless ones,
that lay amongst them, have remained in position, and are still
visible as phantom forms in the Liquor Sanguinis among the
now invisible partly decolourised corpuscles around them.

By the colloid method all the corpuscles of the fugitive group
can be preserved, including the colourless discs, which, however,
as before, are still invisible under ordinary conditions. This
method alone affords a complete answer to the objections
which refer the invisible corpuscles—

    I.—To pressure.

    II.—To violence.

    III.—To the action of saline solutions—*in fact to any
view which ascribes their existence to loss of colour on the part of red,
or slightly coloured corpuscles.* It demonstrates, in the most
absolute manner, that they are corpuscles which have never
possessed enough hæmoglobin to make them visible in the normal
Liquor Sanguinis.

The question of the effects of direct pressure upon the
mammal corpuscle, *i.e.*, of intermittent repeated compressions
applied with great force from above, has also been carefully
investigated, with the result of showing that corpuscles can be
made in this way to give up some of their hæmoglobin, as
evidenced by their becoming paler, and by the liquid becoming
stained. These corpuscles, however, never under any circumstances become less coloured, *i.e.*, whiter than the liquid which
surrounds them. They simply become less visible, because
their own intensity of yellowness is diminished, while that of
the Liquor Sanguinis is increased, and when run into clear,
i.e., unstained Liquor Sanguinis, they are not invisible,
as is the case with the primary disc, but much more visible,
as might *a priori* be readily imagined.

By no amount of pressure or manipulation can corpuscles be
produced in any way corresponding to those I have designated the
invisible colourless discs. All that can be done is to produce a dead
level of colour between the corpuscles and the Liquor Sanguinis.

I have recently discovered a very simple method by which these corpuscles may be displayed, and which has the advantage of not requiring the cover to be strapped down, or, in fact, any plan adopted different to that by which blood is usually examined.

It consists in using flexible covers in the place of the glass ones, such as thin pieces of mica or films of collodion. These covers adapt *themselves* sufficiently close in parts to draw off the Liquor Sanguinis from the corpuscles, and this induces packing in other parts, in which the colourless disc is readily detected. Under these large flexible covers, in some parts, currents are maintained for a considerable time, and ample opportunity is afforded to witness the phenomenon of the impact of the red corpuscles against slightly visible and wholly invisible corpuscles which have become attached to the slide. This phenomenon is, I think, one of the most impressive, for we see the effect of curvilinear indentation produced upon the red corpuscles, while, in the majority of cases, the most searching scrutiny fails to reveal the adhering corpuscle which is giving rise to it.

It is obvious from what has been said, that the detection by this investigator in one instance of hæmoglobin by the aid of the micro-spectroscope is not of the least moment, for as we know that, after a time, in all cases the serum of shed blood becomes perceptibly coloured with hæmoglobin ; we may readily infer that an action of this kind commences to set in immediately the blood is shed, but we may also fairly conclude that all the coloured corpuscles contribute their proper quota to this result. If it be true, as stated by this writer, that some corpuscles yield up *all their colour*, while others are not perceptibly affected, this is at once an evidence of a remarkable difference among the corpuscles themselves, a point by no means to be overlooked.

The experiments with the colloid appear, however, to negative such a view. Again, it must be remembered that, until I pointed out this method of filtering the serum from the corpuscular elements, such an observation was impossible, and we had no means of ascertaining whether hæmoglobin was or was not a normal constituent of the Liquor Sanguinis. There exists an unknown colouring matter in the Liquor Sanguinis which gives rise to its pale yellow tint. This normal pale yellow tint must not be taken as evidence of hæmoglobin.

As methods can now be devised for preventing the passage of hæmoglobin from the corpuscles, and also of removing these bodies from the Liquor Sanguinis, we may confidently hope to be able before long to make out the nature of this colouring matter.*

Objection II.

Method by Raising the Cover.—Under this head my critic writes—"By the method of raising the cover I was enabled, without the use of salt, to make some very good preparations of colourless corpuscles. After fixing with osmic acid, I also stained them. It struck me, however, that as a possible cause of this appearance of these forms of the red corpuscle, the force required to overcome the capillary attraction of the two glass surfaces, with the thin layer of fluid between

---

* My friend, Dr. McMunn, so well and favourably known by his researches on the animal colouring principles, has recently made this matter a subject of investigation, and has favoured me with the following note:—"The points which I have made out about the absorption band of serum are the following: The serum examined was that of the sheep, it gave a band at F. If this were due to luteine, it should have been accompanied by another in violet—it should occupy the position (in the spectrum) of the luteine band—it should be rather intensified by ammonia, or by caustic soda. If, on the other hand, it were due to unoxidised bile pigment, it should be intensified by acids, it should disappear with ammonia, and it should occupy a position in the spectrum nearer to red than the luteine band; the latter is the case, as one can easily prove, therefore the band is *not* due to luteine. The band seen in serous fluids—*e.g.*, peritoneal fluid, pericardial fluid, and that removed by a blister is probably due to luteine; at all events, it does not appear to me at present to be due to an altered bile pigment." Further information may be found on this head in the Author's paper, Poc. Royal Society, No. 208, 1880.

* Cover-glasses are strapped down at one end, forming a kind of hinge, which permits them to be raised by a screw, or otherwise, from the opposite end, and the excess of blood then flows by capillarity to the other, from whence it can be removed by blotting paper, and the adhering corpuscles retained for examination.

them, had not been sufficiently taken into consideration, nor also the probable effect on the more unstable red corpuscles of the sudden withdrawal of the serum."

It is necessary to state, in the first place, that I have never used, nor recommended the use of solution of salt in connection with this method ; indeed, it would be impracticable, on account of crystallisation, but I have employed the method without, and with osmic acid, both in solution and in vapour.

The concluding paragraph of this criticism renders it unnecessary for me to defend this method, or, in fact, any of my methods, for the writer says:—"I may be permitted to say that, after going with care through a series of observations on this subject, I am disposed to believe that the colourless corpuscles, which are undoubtedly seen when the blood is examined by the method of 'isolation' are red corpuscles which have undergone *post-mortem* changes, prior to taking part in the formation of fibrin.'

Here is a free admission that *colourless discs* can be found by one of my methods, and their presence is not referred to the method, but to *post-mortem* change in the blood. If these corpuscles are the result of *post-mortem* change, it is obvious that they cannot be the product of any of the methods employed.

This method of "isolation" is a very important one, because it enables us to understand the relations which various corpuscles hold to the Liquor Sanguinis, and also how they behave when it is withdrawn from them.

By employing it in its simplest form, that is without the use of re-agents, and separating the glasses in the most gentle manner possible, we may obtain groups, such as are seen in the Photographs 9, 10, 11, 12, and 13, Plates III. and IV. As a rule, however, we get only groups of coloured corpuscles, without any of the colourless ones among them. As I have stated, the corpuscles of the "fugitive group" cannot well bear the withdrawal of the Liquor Sanguinis. They owe the integrity of their form entirely to its presence, and spread down upon the glass surface, and are lost, just as oil globules would be, the spherical shape of which had been maintained by submergence in water. The water being removed, the globules spread upon the surface with which they are in contact.

There are certain corpuscles which we may obtain in addition to the mosaic groups of red ones, providing we have our glasses very clean and the right kind of surface, and separate them with the greatest possible care. This is the corpuscle seen in Plate XI., Photograph 59. It is a corpuscle with considerable colour, of a glistening, lustrous, or flickering character ; it has some disposition to spread, and one of its edges may frequently be seen to be laid down. This corpuscle stands between the permanent and fugitive groups, both connecting and dividing them. Generally, when the glasses are separated, these corpuscles are broken up into minute granules. A glimpse of the primary corpuscles is very rarely obtained by this method without the use of osmic acid vapour. When so obtained they are generally supported among red corpuscles.

I consider that I have been able to divide the biconcave discs into three sets, which may be designated the primary, secondary, and tertiary groups. This distinction is founded on the behaviour of these corpuscles upon the withdrawal from them of the sustaining influence of the Liquor Sanguinis.

I.—*The Primary Group* embraces all those corpuscles which melt down on the removal of the Liquor Sanguinis, and, after a time, *even in the Liquor Sanguinis of shed blood.* It includes the whole of the colourless discs, and such of the coloured ones as are less tinted than those of the secondary group.

II.—*The Secondary Group* consists of the lustrous, flickering corpuscles. They are just barely capable of maintaining themselves in the absence of the Liquor Sanguinis. They have considerable colour, in fact, nearly as much as those of the tertiary group.

III.—*The Tertiary Group* includes all corpuscles which do not become lustrous, and which can maintain a distinct outline in the absence of the Liquor Sanguinis.

All these corpuscles occasionally form mixed mosaic groups on the glasses.

Specimens of the corpuscles forming the primary group can only be obtained isolated from those of the secondary and tertiary groups by adopting perfect methods of preservation.

These three classes of corpuscles are all capable of undergoing similar changes, but with different degrees of facility, and, on this account, these changes are commonly seen in the primary or fugitive group only.

These changes are of the nature of *fusion*, of *granulation*, and of *fibrillation*, and groups of each class may be shown in which these changes have occurred or are taking place.

In Photographs 55, 56, 57, 58, 59, Plate XI., we have succeeded in isolating small mosaic groups of the invisible corpuscles, the corpuscles of which have passed into the spherical state. In Photograph 55 their outlines are tolerably distinct, but in Photograph 56 they have partially fused or coalesced, and in Photograph 57 still more completely. In Photograph 58 the masses formed by their coalescence have already commenced to granulate, and would shortly undergo separation into distinct granules, such as are depicted in Photograph 17, Plate IV., and Photograph 45, Plate IX.

Photograph 60, Plate XI., gives an example of the direct conversion of the *colourless discs* into *fibrin* without passing through the stage of granulation. This is a modified mode of action of the process of annulation.*

Photograph 59, Plate XI., shows corpuscles of the secondary and tertiary groups in juxtaposition. It will be seen how incapable the former are of maintaining a distinct outline. It is corpuscles of this class which frequently become granulated, especially under the influence of cold, and thus give rise to the forms observed by Semmer. (*Vide* page 26.)

These primary groups are often to be seen undergoing conversion into fibrin. The corpuscles of these groups are *de facto* fibrin, and the delicate fibres and layers which appear on glass slides are due, first, to the extension of these granu-

---

* *Vide* Section on Fibrin formation.

lations into fibres, or to annulation of the entire corpuscle,
or secondly, to the spreading and laying down of these
corpuscles into films. When blood is completely defibrinated
these corpuscles and their granules entirely disappear, and
can no longer be shown by any of my methods, though
abundance of red corpuscles are still present from which the
colour might be discharged if the methods used could bring
about this effect. One of two things is, therefore, obvious,
either colourless discs have been removed, or the discs which
become colourless when my methods are employed. There
can be no doubt that the former is the true view.

The application of a delicate photo-chemical test, such
as is afforded us by photography, indubitably shows the
existence of a regularly graduated series, from a colourless to
a deep yellow disc. Of these, the colourless and the more
faintly-tinted ones range themselves together on the unstable
or fibrin side, and the more strongly-tinted on the stable
or permanently corpuscular side; in other words, the stability
of the blood corpuscle is directly proportionate to its degree
of cruorisation, and the flickering or diffused edged corpuscles
mark the point at which the biconcave discs become converted
into fibrin when the blood is shed.

It must not, however, be supposed for one moment that
we have in these discs to deal with a difference of *colour*
alone.

The corpuscles which constitute the "fugitive group" differ
not only in colour but also in adhesiveness, in specific gravity,
in liquidity, in their relation to stains, in their tendency to
granulation, in their behaviour with re-agents, &c.

Although, as before explained, many of the corpuscles of
the "fugitive group" possess some colour, yet on the whole the
colourless disc may be taken as the type of their behaviour;
the red disc as the type of the behaviour of such as are more
strongly-tinted than the diffused edged one. It will be simply
necessary, therefore, to describe here the general properties of
the *colourless*, or as I sometimes designate it, the *primary* disc.

Colour.—The primary discs are of the same tint as the ordinary white corpuscles when these have been rendered smooth and free from granulation, *i.e.*, that of the background. The rosette masses, which they form by their adhesion to each other, are also of the same colour as similar masses of adhering white corpuscles.

Adhesiveness.—They are more prone to coalesce with each other and with the *white corpuscle*, but show little tendency to associate themselves with the red. They have also a great disposition to adhere to air-bubbles and foreign matters introduced into the blood, while the red discs have little or none.

Specific Gravity.—Like the white corpuscle, they are lighter than the red, and have a tendency constantly to rise to the surface of the blood, consequently the largest numbers are always seen to attach themselves to the upper glass, in preference to the lower, and especially if time is allowed them to rise. This, no doubt, has something to do with the *buffy coat.*

Liquidity.—In this respect they differ much from the red corpuscles. When in their natural state in the blood they are exceedingly liquid, and have no power to retain their form, being modified from moment to moment by currents.

It is not till they have formed adhesions to the glass, or have been hardened by re-agents, that their true form becomes obvious. So great is their natural liquidity, that they may be often seen to give off under the influence of capillarity, finger-like processes which lie between the more permanent corpuscles, *vide* Photographs 15 and 16, Plate IV. In its natural state in the blood there is reason to think that this corpuscle is far more plastic and yielding, but less elastic than the red corpuscle.

Granulation.—The red corpuscles rarely undergo granulation, but these can scarcely be prevented doing so. In this respect they are like the ordinary white corpuscle, to which body they in fact assimilate in *all* their properties. These granules sometimes result from the breaking up of single corpuscles, and at others from the breaking up of groups or fused masses of them, *vide* Photographs 12, 17, 18, and 45, Plates III., IV., and IX., and Photograph 58, Plate XI.

These are the bodies which M. Hayem has described as the *hæmatoblasts* of mammalian blood. The body, however, which in reality corresponds to the delicate corpuscles he has discovered in the blood of the ovipara, is the *invisible colourless disc.**

Relation to Stains.—I have already stated in my former paper that these corpuscles when in the Liquor Sanguinis stain with carmine, and the red ones with aniline. I have found more recently that the colourless discs may be readily stained by a weak

---

* Dr. Noel Gueneau de Mussy, the distinguished honorary physician to the Hotel Dieu, has done me the great honour to reproduce my work on the blood with remarkable exactness, and with a degree of lucidity which shows that he has taken considerable pains to acquaint himself thoroughly with my views. That he has much confidence in their general trustworthiness, may be inferred from the following passage :—

" Cette accumulation d'inductions, d'observations, d'expériences si nombreuses et si ingénieusement variées, ne me paraît guère permettre de contester l'existence de ces nouveaux corpuscles ou disques incolores."

There are, however, one or two points which seem to require from me some explanation. These refer to the relations which subsist between my work and that of M. Hayem, and are as follows :—

" Il combat l'opinion de M. Wharton Jones, qui veut trouver dans les noyaux des leucocytes l'origin des hémoglobules. La théorie, dont il est le défenseur, et qui placé cette origine dans les globules de la lymphe, avait été déjà entrevue ou soutenue par plusieurs physiologistes, entre autres par Kölliker et Huxley.* Elle est en opposition avec celle qui a été proposée dans ces derniers temps par notre savant confrère M. Hayem, qui fait naître les hématies de globules rudimentaires auxquels il donne le nom d'hématoblastes. Ces petits corps seraient, selon M. Norris, ceux qui ont été décrits par Beale sous le nom de *bioplasts* et il les considère comme des granulations produites par la rupture des disques incolores. Je me demande si M. Norris a bien compris la description de M. Hayem, qui ne me semble pas se prêter à cette interprétation.

" Sans doute, il y a entre les disques lymphoïdes de M. Norris et les hématoblastes de M. Hayem des différences essentielles de forme, de

* Dr. de Mussy is in error here. Professor Huxley supports the views of Wharton Jones.

solution of aniline blue, in three-quarter per cent. solution of common salt. The blood should be allowed to run in by capillarity at one end of a large mica cover, and the stain should be so applied as to be drawn in after it. If the stain is not too strong it will tint the invisible corpuscles without affecting the red ones.*

volume, d'origine, mais la description qu'en donnent ces auteurs laisse voir aussi entre eux certaines analogies dont la plus saillante est le rôle que tous deux font jouer à ces corpuscles dans l'origine de la fibrine et la coagulation du sang. De nouvelles recherches me paraissent nécessaires pour décider lequel de ces deux observateurs a le mieux vu, et pour éclairer l'origine des hémoglobules, dont la théorie de M. Norris donne, il faut en convenir, une séduisante explication.

"Quand bien même les recherches de M. Hayem sur les hémato-blastes et sur leur rôle dans la genèse des hématies recevraient la sanction des observations ultérieurs, l'existence des corpuscules incolores de M. Norris n'en serait pas ébranlée, leurs connexions avec la fibrine pourraient subsister."

I desire, in the first place, to bear my most unreserved testimony to the originality and importance of M. Hayem's investigations. I regard them as the most profound researches which have been made in the blood for many years. Very little recognition has yet been accorded to M. Hayem by English physiologists, for the reason that very few have worked sufficiently at the blood to enable them to form an opinion upon the merits of so obscure and difficult a research, and also because English Physiology is not yet educated up to " a morphological fibrin." Although M. Hayem and myself have laboured independently of each other, the main conclusions at which we have arrived are singularly in unison, more particularly those which refer the formation of fibrin and the coagulation of the blood to such formed elements of the blood as are destined in the ordinary course of things to become the red corpuscles.

Neither of us deny that the ordinary white corpuscle may contribute in a small degree to fibrin formation, but we are both agreed that it is not the normal progenitor of the red disc.

Up to this point our researches are mutually supporting.

M. Hayem has very carefully described certain minute bodies long known to exist in mammalian blood, and has shown that these bodies are concerned in the formation of fibrin. He had already shown that the bodies which form fibrin in the ovipara were the early forms of the red blood corpuscle, and he has, therefore, naturally inferred that these variable elements in mammalian blood must be the early stages of the red blood disc. These bodies have some colour, and are but a fraction of the size of the red disc, therefore the further assumption that they grow and attain more colour has to be made. On the other

_____

Vide section on staining.

OBJECTION III.

By altering the refractive index of the Liquor Sanguinis.[*]—
The ground having been cleared by the previous discussion, we
are now in a position to consider more profitably the objections
which have been made to the use of a saturated solution of salt
as a means of displaying the new corpuscles by altering the
refractive index. I have already referred to the grave mistake
made by my critic in attributing to me the use of salt in all
my processes instead of in two only. (*Vide* page 42.)

On this method my critic remarks—"As to the use of
a saturated solution of chloride of sodium, I conclusively proved,
by repeated experiments with the instruments used for the
enumeration of the blood corpuscles, that a saturated solution
of salt causes about one-third of the red corpuscles to discharge
their hæmoglobin, and to become clear, colourless, ghost-like
corpuscles."

The fact that a certain number of the paler red discs
disappeared in a concentrated saline solution was very well
known to me, for when I decided to use a saturated solution

---

hand, I find that the fugitive corpuscles, which are the same size
as the red disc, are in various ways resolved into fibrin, and that
in the process they often break up and become altered into the various
minuter forms depicted by M. Hayem in "The Archives of Physiology,"
pages 731 and 732. If this able investigator will trace these forms still
further back, he will find that they are the disintegrations variously
modified of partially coloured and colourless biconcave discs, and that
it is these discs which are the true analogues of his oviparous
hematoblasts. *Vide* section entitled "An Examination of the Researches
of M. Hayem on the Development of Mammalian Blood."

* If we place upon the tip of the finger a minute drop of saturated
solution of salt, and prick through it so that the blood may flow directly
into the saline solution, the refractive power of the Liquor Sanguinis is
modified, and it is found that if we run this mixture of salt and blood
between glasses prepared according to the packing method before
described, we can then see the *outlines* of the colourless discs, and
the clear spaces which have hitherto been supposed to consist of Liquor
Sanguinis only, are observed to teem with these discs. After a time
specimens thus prepared become tinted with hæmatin, and this
stains the edges of the colourless discs, and renders them still more
apparent.

of salt, I first carefully went into the question of its effects
upon the *visible* corpuscles.

This solution has a certain limited power of dissolving hæmoglobin, but it is an action that is soon satisfied, for I have kept corpuscles in contact with it for a
week in hermetically-sealed tubes without much change
either in the colour of the corpuscles or of the solution.
When blood is added to such a solution, there is no doubt
that all the corpuscles give off a minute quantity of colouring
matter. This action, while it would tend to render all the
corpuscles a shade paler, also levels up the surrounding liquid
to their colour, and so certain of the paler corpuscles disappear,
not simply because they have lost colour, but also because the
solution has gained it. Even when water has been added to
blood, so as to render the corpuscles almost indistinguishable,
we have only to get them out of the coloured serum to see that
they have individually lost but little colour, and that the action
has been mainly one of levelling up. Such ghost - like
corpuscles placed again in uncoloured serum are still seen to
possess a pale yellow tint.

After I had satisfied myself of the existence of corpuscles
that could not be seen, it did not seem to me a matter of
much importance that a few pale visible corpuscles should be
by this levelling-up action added to the invisible set. I adopted
this method merely to prove what might *a priori* be expected, that
the colourless discs would be brought into view by alteration of
the refractive index of the liquid in which they lie, and also to
preserve them, so that they might be stained, for after I had
acquired the knowledge that these corpuscles were the *fibrin
factors*, it occurred to me that, as neutral saline solutions
prevented coagulation, they might do so by hindering physical
changes in these corpuscles, and on examination I found this
view to be correct. My critic proceeds :—" I therefore sought
for other means of changing the refractive index of the
serum, such as fixing the blood immediately it is shed, with a
two per cent. solution of osmic acid, and by diluting the blood

thus fixed by large quantities of distilled water. In such olu-
tions in which the corpuscles have been preserved in eir
normal condition, and the serum so diluted that its refracuve
index is changed, I did *not* succeed in obtaining any transparent corpuscles, examine in thin layers, in the ordinary
manner."

On this I would remark that I should not expect to do
much in the way of altering the refractive index by that
method, because the blood liquid contains only ¾ per cent. of
chloride of sodium. It would not by any means produce such
a difference in density as is effected by using a saturated
solution of salt, and, therefore, if we did not bring into view the
colourless discs by this method, we should have no right to infer
that they did not exist. My critic considers that the osmic acid
prevents the corpuscles losing their colour; that, in fact,
it fixes the colour, and therefore no colourless corpuscles
are present, because none have been produced. I have
repeated this experiment by pricking through a drop of osmic
acid on the tip of the finger, bringing the blood directly
into contact with this fixing agency, mixing quickly and
perfectly. I find that, if this mixture is run under glasses
prepared for the packing method, the colourless corpuscles
are present, as usual, having, I think, a little more distinctness of outline. Having thus ascertained that the osmic
acid so applied did not interfere with their presence, I
proceeded in a second experiment to dilute with water,
and here, too, I found colourless discs present in abundance.
The failure of my critic to see them must probably have arisen
from the fact that there was great paucity of corpuscles, with
excess of liquid. If we run the whole of the preparation under a
large mica cover, taking care that when it is laid down there is
still ample space for liquid to filter off, we shall find, after a
time, that the colourless discs will come into view in the
free spaces between the red corpuscles. In about an hour
they may be seen to greatest perfection, for by that time the
layer of liquid above them has become sufficiently thin.

They may, however, be seen in some parts as soon as the preparation is made.

The statement, therefore, that "if the refractive index of the serum be changed by means which are conservative and not destructive of the integrity of the red corpuscles, the transparent or third corpuscle cannot be found by any of the ordinary methods of observation," I cannot accept as correct.

In conclusion the principal reasons why the *invisible colourless discs* cannot be regarded as *decolourised red discs* may be thus briefly summarised:—

1.—In the methods of "impact," of "packing," and of "isolation," neither pressure, compression, nor violence are present, and if they were, they could not convert coloured into colourless discs.

2.—These colourless discs exist under conditions in which the decolourisation of red corpuscles cannot occur (colloid method.)

3.—The colourless discs can be stained by a preparation (aniline blue) which contains that proportion of salt ($\frac{3}{4}$ per cent.) affirmed by others to be consistent with the retention of colour by the corpuscles. This preparation, when of the proper strength, does not stain the red corpuscles.

4.—They are present, as usual, when the blood is introduced directly from an artery into a $\frac{3}{4}$ per cent. solution of salt fully saturated with hæmoglobin.

5.—They have neither the physical nor chemical constitution of decolourised red discs, but of *lymph or gland corpuscles.*

6.—The disintegrative changes which take place in these corpuscles give rise to the formation of fibrin in the blood, and the fibrin which is formed in the lymph has its origin in similar changes in the most advanced gland corpuscles.

SECTION III.

DIFFICULTIES of a very special character associate themselves
with the attempt to bring out the *colourless and the intermediate
discs* of the blood by means of stains. Had this not been so it
is probable that these corpuscles would have been long since
discovered during the study of the white blood corpuscles by
these methods.

The corpuscles of the " Fugitive Group " have a moderate affinity for certain stains, which diminishes in the ratio
in which they assume colour. As the biconcave discs of the
blood represent one continuous graduated series, without break,
it is not to be expected that the staining will occur in any other
than a graduated fashion, that is to say, there will be no sharp
line of demarcation, on one side of which we may place discs
which become stained, and on the other discs which do not
stain at all ; but if we take the extremes, viz., on the one hand,
the invisible corpuscle which takes the stain strongly, and on
the other, the fully coloured red disc which refuses to take it at
all, we can readily place the intermediately stained corpuscles
in their proper position in the graduated series; thus, if we use
aniline blue, which is an excellent stain for this purpose, we
shall find that the invisible corpuscles stain of a deep blue tint,
while those corpuscles which have acquired a little hæmoglobin
assume a greenish tint, and those which have much, fail to stain
at all.

These differences in relation to the dye are found in the
varying chemical composition of the corpuscles themselves. A
body consisting mainly of paraglobin is undergoing gradual
conversion into hæmoglobin, the former has an affinity for the
stain, while the latter has none, hence the corpuscle stains in
the inverse ratio of its degree of colour.

The green tint presented by some of the intermediate corpuscles is simply due to the combination of the blue aniline, and the yellow hæmoglobin.

The staining processes may be divided into the wet and the dry. In the former it is sought to stain the corpuscles as they lie in the Liquor Sanguinis, and in the latter after withdrawal from this fluid. Each condition requires its own peculiar treatment.

Wet Method. Staining the corpuscles as they lie in the Liquor Sanguinis is rendered difficult by two circumstances. 1 The Liquor Sanguinis possesses a singular property of preventing the staining of the fugitive corpuscles and also of discharging the colour from them after they have already become stained. This renders it necessary to use a strong staining fluid and such mechanical arrangements as will tend to bring the whole of the corpuscles into contact with the staining fluid. 2 Strong aniline blue solution breaks up the colourless discs into granules.

If we take a drachm of a ¾ per cent. solution of salt, and add to it 2½ or 3 grains of aniline blue, dissolve and filter through flannel or lint, we have a fluid which will stain the corpuscles of the "Fugitive Group"—the invisible ones blue and the others bluish green, or greenish blue, according to their degree of colour.

This preparation requires, however, to be used in a particular way.

Let a cover-glass be attached to the slide, and a little blood allowed to run under, sufficient to fill about a fourth part of the space, and then let the stain be introduced at the same spot. As it passes in it will sweep the mass of the corpuscles before it, but many of the younger corpuscles have already attached themselves to the slide and cover, and will be found to be deeply stained.

When we try to use this stain in other ways, say by adding it to the blood, either by pricking the finger through a small drop of it, or by placing a drop of blood and a drop of staining-fluid in juxtaposition on the slide and allowing them to mix when the cover is lowered down upon them, we get quite a different state of things—we see only large masses of stained granules.

These owe their origin to the breaking up of the corpuscles, which we desire to stain and display.

This is due partly to the fact that time had not been allowed them (as in the first case) to become attached to the glass, prior to their being attacked by a more limpid, that is a less colloidal liquid, than their natural plasma, and partly to the granulating action of the aniline.

With a view to obviate this, staining fluids were made, containing different quantities of albumen, but it was found that although this substance prevented granulation it did not prevent the invisible corpuscles from adhering together in smooth masses, so that now we had corpuscular, instead of granular masses.

It was clear that, for successful staining, something must be done to retain the corpuscles in the same non-adhesive state which they possess when in the blood current. Many substances were tried with this object, but the best of these proved to be ordinary white cane sugar. When this substance is used in proper quantity it completely holds in check the adhesiveness of all the corpuscles.

Various ways suggest themselves for the use of this re-agent, but they are by no means equally good. The method which I advise, and which 1 have found to be most successful after many trials, is to reduce the sugar, in the first place, to an impalpable powder—taking a portion of this powder, place it on the tip of the finger, and prick through it, so that the blood may come in contact with the sugar immediately it leaves the vessel—blood may thus be converted into a syrup of any degree of consistency; this blood syrup may be run under a cover glass at one side, and the saline aniline may be run under at the opposite side. When the blood syrup and the stain meet each other the invisible corpuscles will be seen to stain *at the line of contact.*

In order that these corpuscles should be quickly stained, it appears to be necessary that they should come into contact with the staining fluid while it is at its full strength, or at least before it has become much diluted by the Liquor Sanguinis. It might be thought that to meet this difficulty it would be only necessary to increase the strength of the staining fluid. This is not so, for if we tinge the whole of the blood liquid deeply, the corpuscles stain, but they do not become properly visible, because they still lie in a liquid, having pretty nearly, if not quite, the same colour as themselves. What is really wanted is that they should stain, and then leave the stained liquid for a lighter one ; besides, if we use the staining fluid too strong, there is an increasing liability to stain also the red corpuscles.

The following is a very good method for getting a general idea of the varying degrees of staining power possessed by the corpuscles of the " Fugitive Group," and for shewing that the permanent group has little or no disposition to stain : run under a cover glass sufficient, $\frac{3}{4}$ per cent., of salt solution to fill the space quite full, then place on the end of the finger a little very finely powered cane sugar, prick through it and squeeze the finger so that a small drop of blood may exude into the sugar, mix well with the needle and then add a drop of about the same volume of 10 per cent. aniline blue in $\frac{3}{4}$ per cent. of salt solution, mix as before, and then transfer to one edge of the cover glass, and examine under the microscope the portion of blood which runs into the clear saline solution at the edge of the cover—saturated solution of cane sugar under the cover glass is an excellent variation of this experiment—in it the stained corpuscles retain their colour for a longer period.

Glasses arranged as for the packing method are exceedingly useful to show the effects of staining. The blood may be mixed as usual with a small portion of sugar on the end of the finger, and to this may be added a small quantity of a $\frac{3}{4}$ per cent. saline solution of aniline blue (10 per cent.), and after well mixing with the needle the preparation may be run under the cover glass. The colourless discs will first be seen in the unstained state in contrast with the deep blue surrounding, subsequently they will stain and become lost to view, and will afterwards re-appear as they become more deeply stained than the tinted Liquor Sanguinis in which they lie. This method has the advantage of proving that the corpuscles which become stained are the self-same corpuscles which show themselves as clear circular spaces when the packing method is employed.

Mica covers may also be employed, time being allowed for the corpuscles to stain and pack. In all other respects we proceed as before.

Dry Method. Whenever red blood corpuscles are allowed to dry upon the slide, *e.g.* after being picked out of the blood by the ' method of isolation,' or after being spread by means of a glass rod, it is impossible to subsequently add to them aqueous solutions of any kind, or even Liquor Sanguinis of fresh blood without causing solution of their hæmoglobin. In such cases, therefore, it is necessary before applying any staining fluid to thoroughly fix the corpuscles by subjecting them, say for five minutes, to the vapours arising from a saturated solution of osmic acid. After this treatment the staining fluids may be used with impunity. All dry specimens fixed with osmic acid require that the strength of the aniline blue shall not be less than 2 per cent., and that it shall be applied for about two minutes. It is necessary to add that the colour must not be washed off, but removed by capillarity as in the ' isolation method.'

In no case must the blood be allowed to dry between the osmic acid and the application of the staining fluid, as this renders the specimen impermeable to liquid, and therefore incapable of absorbing the stain.

Many of the corpuscles which stain would of course be visible in the absence of the staining, but others, *i.e.*, the youngest ones, are neither visible or capable of being photographed; indeed, some of the corpuscles which stain of a deep blue tint, and are therefore now exceedingly patent to the eye, are still incapable of being photographed, while others that have stained nearly as much may be shown by photography, owing to the fact that they possess a small amount of hæmogloblin. We can, therefore, by this means, distinguish among the corpuscles which stain those which contain a little and those which are free from hæmogoblin. This matter has been made the subject of special investigation.

An Examination of the Researches of M. Hayem on the Development of Mammalian Blood.

In the year 1877 M. Hayem presented to the Academy of Sciences, and to the Society of Biology, a new research upon the blood, under the title of " Récherchés sur l'evolution des Hématies dans le sang de l'homme et des vertebrates."

This was subsequently published in detail in 1878, in the Archives of Physiology.*

It is the object of this paper to examine into the accuracy of these views, so far as they relate to mammalian blood, and also to point out the connections they have with my own published researches upon this question.

In doing this the better plan would seem to be to begin by a careful statement of the views of M. Hayem, allowing him to give his own description of the bodies to which he specially draws attention, and subsequently to examine the methods which he employs to display them, and finally to set forth the conclusions at which I have myself arrived after traversing the same ground.

In the first place, this author affirms that there are in the blood of all vertebrate animals small elements which are neither white nor red corpuscles. These elements " may," he says, " be styled the germs of the red corpuscles, and are the youngest forms of them." He proposes to call these elements *hæmatoblasts*. It is freely admitted that these bodies are not new elements of the blood, and that the facts which are now adduced respecting them are not all new, and that many authors have given descriptions which may be considered with more or less probability to refer to them.

M. Hayem does not undertake to discuss the origin of these elements, nor the different forms which they may assume

* Archives de Physiologie, normale et Pathologique, publiées par Messieurs Brown Sequard, Charcot, Vulpian.

during the several periods of the evolution of their being—nor the manner in which they are formed, nor the organs in which they arise.

Taking the animal at birth, he devotes himself exclusively to the examination of its blood, in order to arrive at its exact anatomical constitution.

It is important to remark that the bodies to which he desires to draw attention are *visible* elements of the blood which may be seen as readily as the white and red corpuscles, without the addition of any re-agents or the adoption of any arrangements differing from those usually employed.

The various methods suggested and explained are directed mainly to the preservation of those bodies for a lengthened period—to the maintenance of their true form, and to the retention of their colour.

Thus, we are told, that in order to properly examine the hæmatoblasts, it is absolutely requisite to take the blood as soon as and in the condition in which it leaves the vessels.

The following is the method of procedure :—" Clean the glass-plates with alcohol or ether; dry them carefully, then fix the cover-glass to the slide, by dropping at each corner of the former a spot of melted paraffin. A capillary space is thus obtained to receive the blood.

" Place the slide upon the stage of the microscope, so that the elements of the blood may be observed immediately they penetrate by capillarity between the two glass plates.

"It is necessary to employ considerable magnifying power,* and to cause the blood to drop upon the edge of the cover-glass at the very instant at which it is squeezed from the pulp of the finger.

"As soon as the blood arrives at the capillary space, it rushes in with great velocity, and the various elements may be remarked quickly passing and rolling about.

" At certain points, which are easily distinguished, the blood

---

* Or two or three—ob. 5 Nachet.

current is slower, and here the elements sought may be perceived. Thus, among the red and white corpuscles may be remarked other *very small ones*, which, at first sight, seem to be very delicate and *pale red* corpuscles.

"Almost as soon as seen they begin to change, throw out points, adhere to the glass, double up, grow pale, consequent upon the loss of the whole or a portion of their hæmoglobin, and tend to join themselves to other corpuscles which they encounter, so as to form a mass.

"Sometimes they arrest a passing red corpuscle, which adheres to a point of their circumference, whilst the current tends to drag it along, causing it to assume the shape of a pear. After some time the red corpuscle disengages itself and passes on to contribute to the formation of rouleaux, and the small elements remain isolated, or form wreaths or groups with each other.

" By this time they are much changed, and almost irrecognisable, but their presence has been now ascertained, and their transformations have been followed, and the fact established that in addition to the red and white corpuscles, the blood contains some peculiar elements which change their appearance very rapidly."

In order to permit of these bodies being investigated with greater deliberation various methods are proposed, which are held to be calculated to delay or prevent the spontaneous changes to which they are so prone.

Thus, the above experiment may be repeated at a temperature of 0° Cent. or lower.

All M. Hayem's experiments were made in the open air during the winter, so that everything was cooled down to or below the zero point.

" It is easy, however, by the use of the cold stage, (as suggested by M. Hayem,) to repeat these experiments at ordinary temperatures. M. Hayem has obtained the best results at .1° Cent., but good observations may be made at from 1° to 1°·5 Cent.   In this case the " red corpuscles

arrange themselves in rouleaux as at ordinary temperatures
and in the spaces between the rolls there may be seen small
very delicate bodies, isolated or gathered into groups of two,
three, four, or five, seldom more."

"These small bodies are remarkably clear in outline,
although very delicate and thin ; the majority of them are
obviously discoid, biconcave, and slightly coloured, others are
elongated, and have a kind of pedicle more or less long."

Other methods have been employed, such as the use of
osmic acid, of bichloride of mercury, and of spreading and
and drying the blood.

From the results obtained by these various methods the
author feels himself able to give the following definitions of
these bodies :—

"*Definition of so-called hæmatoblasts* (as seen at the temperature of room.)—Very little corpuscles, resembling at first little
red globules, very delicate and pale ; almost as soon as seen
they begin to change, throw out spikes, adhere to the glass,
double up, grow pale by loss of the whole or a portion of their
hæmoglobin, attach themselves to other corpuscles which they
encounter to form a mass, remain isolated, cr form wreaths
or groups, and after a time fade away and become
irrecognisable. (As seen at the temperature of 1°:)—Small,
very delicate bodies, in groups of two, three, four, or five,
remarkably clear in outline, although very delicate and thin,
the majority obviously discoid and biconcave, slightly coloured,
others elongated, and having a kind of pedicle more or less
long ; some show themselves as staffs, (bâtonettes,) or grains of
rice. They are mostly elements seen edgeways, and have consequently greater refractive power, and are surrounded by a
deeper shadow than those which are seen flat. They are perfectly homogeneous, and have smooth surfaces, a colloid look,
and nearly always a perceptible greenish or yellowish tint, so that
their substance resembles that of a red corpuscle, but slightly
coloured. The smallest, which have no granulations, are sometimes colourless, but bear no resemblance to the white corpuscles. ˙

In respect to size they are stated to have a diameter of 1·5 to 3-1,000ths of a millimetre.[*]

M. Hayem gives a further description of these bodies, which is worthy of notice from the fact that it recognises that they are composed of two distinct substances, and that they possess definite relations to fibrin formation.

He says, " using a magnifying power of from 500 to 800 diameters, it is easily seen that these bodies are angular, and that they present two parts, more or less clearly divided, the one being the circumference, which is greyish, or finely granulous, the other being the central part, which is corpuscular, glossy, ovoid, and considerably refractive. Their first alteration consists of a kind of contraction, which renders them more brilliant and sparkling, and causes the exuding around them of a peculiar kind of matter."

"The substance which is thus exuded by the hæmatoblasts is very viscous, and this explains satisfactorily the formation of small masses. When first seen, the hæmatoblasts, which are united together, have the form of small angular grains, or little stars, and form frequently garlands, in which each grain can be picked out. Then these small corpuscles appear to attract each other strongly, and the viscous substance which surrounds them tends to form a common mass, in which the constituent elements place themselves and become mixed. From the edge of this small mass a great number of fine prolongations may jut out."

I shall now endeavour in the light of my own researches to identify these corpuscles of Hayem, and to ascertain whether or not they are bodies with which I am acquainted, and if so, ask myself in what position I have already placed them in my scheme of the development of the blood, and also what histo-chemical properties I have ascribed to them.

The formed elements of shed blood which can now be demonstrated to exist, in addition to the well-known red and

---

[*] From about $\frac{1}{16800}$ to $\frac{1}{8000}$ of an inch.

white corpuscles, and the fibrin, are :—

I.—The invisible corpuscles or colourless discs.

II.—The pale visible corpuscles which occupy an intermediate position, and fill up the gap between the former and the red disc.

III.—These colourless and pale intermediate corpuscles in incipient states of granulation presenting the appearance of small groups of adhering granules.

IV.—Masses formed by the coalescence of such granulating corpuscles.

V.—Isolated or separate granules, resulting from the complete breaking up of these corpuscles and the masses which they form.

VI.—Minute granules due to the disintegration of the ordinary white corpuscles.

Among these structures I have no difficulty in identifying the elements which M. Hayem has mistaken for the germs of the red corpuscle.

They are, without doubt, the modified forms of the younger discs, which I have described under the collective term of the " Fugitive Group," and they have consequently no more claim to be regarded as the germs of the red corpuscle than this derivation may confer upon them.

The task of identifying these bodies has been rendered very easy, by the minute and accurate descriptions which M. Hayem has given of their size, colour, chemical and physical properties, and of the relations which they bear to fibrin. Before, however, proceeding further, it may perhaps be worth while to compare the general properties of these bodies with those which I have ascribed to the intermediate corpuscles of the " Fugitive Group."

I.—They are described as possessing colour, but are nevertheless paler, that is, less highly coloured than the red corpuscles.

II.—They exhibit adhesiveness, and by this means attach themselves to the glass, and also form groups with each other.

III.—They are often described as possessing a liquid, lustrous, highly refractive appearance.

IV.—They are very fragile, quickly undergo change, and suffer both in form and conspicuity.

V.—They associate themselves in certain definite ways with fibrin formation.

VI.—They have certain well-marked relation to stains.*

By reference to my previous papers, the reader will see that these are precisely the chemical and physical properties which my investigation has brought out in respect to the " Fugitive Group" of discs, some of which qualities are in fact implied in the designation. There is another property, on which I have laid particular stress, and which explains the varieties of form and size under which bodies possessing these properties appear in shed blood, this is the extreme tendency which the intermediate corpuscles of the "Fugitive Group" have to present themselves in modified and granulated forms. It is not, however, my intention to rest on these general analogies, but to give actual examples of the disintegrations of the corpuscles of the "Fugitive Group," and to show that these are the bodies described by M. Hayem. The proof will consist, firstly, in showing that the corpuscles of this group break up into bodies, having the size, form, and colour of M. Hayem's hæmatoblasts, and secondly, in demonstrating that if measures be taken which prevent the breaking up of these corpuscles no such bodies can be found in the blood, thus affording both positive and negative evidence of their fragmentary character.

I shall then, in the first place, confine myself to tracking the visible corpuscles of the " Fugitive Group" through their various modifications and transformations, and, in order to do this methodically, I propose to fix the attention of the reader upon a group of corpuscles withdrawn from fresh blood, by the " method of isolation," in combination with the use of the vapour of osmic acid—*vide* Plate XII., Photograph 61. In this group

Vide Section on Staining.

there are to be found examples of every kind of blood disc, ranging from the colourless or invisible to the full red corpuscle.

We have here, in a limited space, the means of studying some of the peculiarities of all these discs, and we observe that even here, preserved as they are by artificial means calculated to fix them at once in the condition in which they happen to be at the moment they were withdrawn from the blood, great differences exist. Thus we notice that the youngest of the series actually lie among the others very much as a simple liquid would do, sending in finger-like processes between the more stable corpuscles.

From this we learn that the most primary corpuscles are not only as colourless as the Liquor Sanguinis, but that they are in the normal condition of the blood, nearly as liquid, and this, no doubt, furnishes another reason why they have not been earlier observed.

When from these we pass to such as have a slight trace of colour, we note that with this accession of colour there is likewise a corresponding power of retaining form ; but even these corpuscles are seen to be large in comparison with those still more coloured, owing to their greater disposition to spread out upon any surface with which they may come into contact. The most coloured have, therefore, the smallest diameter, because their disposition to do this is least. After blood is shed there appears to be also a disposition on the part of even the young coloured corpuscles to increase in size, probably from the imbibition of liquid. *vide* Photograph 89, Plate XIV.

In groups such as those shown in Photograph 61, we have, therefore, liquidity, size, and colour to guide us in our conclusions as to the true position in point of age of the corpuscles which constitute them, and we are thus enabled, without difficulty, to place these corpuscles in their proper serial order, the one extreme of which is represented by a liquid colourless disc, and the other by a comparatively stable, highly coloured one. Between these two extremes lie the series of corpuscles, which, for convenience, I have termed " the intermediates," and these must be also held to occupy an intermediate position in respect to the whole of their physical and chemical properties. Those which are more closely related to the invisible corpuscles will approximate to them in property, and those which are nearest the more highly coloured ones to them.

I have stated before that the so-called hæmatoblasts of M. Hayem, so far from being the germs of the red corpuscle, are, in reality, modifications and disintegrations of its younger forms, that, in fact, they are connected, not with the integrity and life of the corpuscle, but with the changes which accompany its dissolution and disappearance as a corpuscular structure, changes concurrent with the shedding of the blood.

It will now be my duty to show how the intermediate corpuscles give rise to these bodies, and to describe and illustrate the various stages of the changes which these corpuscles undergo before they reach that state of complete granulation which has been so frequently recognised in the blood, and from which point they have been so carefully and minutely described by M. Hayem.

In every experiment where fresh blood is submitted to examination, either at ordinary temperatures, at blood heat, by means of the warm stage, or at the freezing point or below, there appear, in addition to the red and white corpuscles, a number of variously coloured groups apparently formed of minute granules, and also isolated granules of the same kind as those which form the groups. Such isolated granules, and granule-groups, may be seen in Photographs 45, 64, 65, 68, 70, 71, 72, 75, 79, 80, 81, 87, 90, 91, 92, and 93.

It is with the nature and origin of these granules and granule groups (real and apparent,) that we have now to concern ourselves. M. Hayem's conception is that they are distinct bodies, which, being mutually adhesive, have become united with each other into little groups. This view is not, however, invariably correct, for I find that some of the smaller apparent groups, which are seen when the blood is first inspected.

are, in reality, corpuscles in the act of undergoing granular disintegration. This is readily seen by removing from them the Liquor Sanguinis by the method of raising the cover—(isolation.) These granulating corpuscles, having already adhered to the glass slide and cover glass, are not removed with the bulk of the blood, and we are then able to ascertain clearly that they are in reality adhering corpuscles, which are undergoing two distinct forms of disintegration, which I propose to distinguish as the rosette and granule form. In the first case bodies are produced having a white border, formed of smooth granules, and a centre, which has a greenish, aqueous, lustrous look, and we become aware at once, both from their size and their general appearance, that we have under our eyes young blood discs in a modified state, and we are also aware, too, that we are witnessing the manner in which *some* of the free granules seen in the blood are formed.

By the use of the warm stage we appear to facilitate this change, for we find that the white borders of the specimens so prepared are more spiked and star-like.—*Vide* Photographs 67 and 69, Plate XII.

As the green, lustrous centres photograph of a rather dark tint, we know that the change has occurred to corpuscles possessing some colour. On the other hand, when an ice-cold temperature is used, the action seems to be more confined to less advanced corpuscles, that is, to those possessing less colour, or the more primary ones.—*Vide* Photographs 62 and 63, Plate XII., and their descriptions.

The central, greenish-looking, lustrous part constitutes a sort of centre or focus, from which coloured fibres frequently radiate in every direction.

Broken portions of these corpuscles appear to behave in the same manner as entire ones, the fragments also becoming surrounded by a white exudation border.—*Vide* Photograph 68, Plate XII., and description.

As a consequence, these centres are very variable in size, some of them being produced by the green, lustrous portion of entire corpuscles; others by that of several corpuscles fused into a mass; and others by that of fragmentary portions only. It is by the changes which occur in the centres of these corpuscles and of their granules that the radiating kind of fibrin is produced. It must be borne in mind, however, that this is only one mode in which fibrin is formed.

I have previously pointed out that the corpuscles of the glands and of the spleen exude, under certain circumstances, a content matter which is more colourless and liquid than their exterior. This material frequently arranges itself around the contracted pellicle which previously contained it, and thus a nucleated cell is simulated. At other times the content matter gets entirely free of the pellicle, and it then swims freely about, presenting the appearance of a homogeneous liquid globule. This observation seems to afford the key to several similar changes which occur in connection with the " Fugitive Group " of discs. As these discs are now to be regarded in the light of gland discs in a further stage of development, it may reasonably be expected that they will continue to agree with these in at least some of their properties.

The green lustrous corpuscles, which become the centres or foci of radiating fibres, appear to present us with the same phenomenon in a slightly modified form. When intact, they present themselves as lustrous, liquid, colloid, greenish corpuscles, having no perceptible difference of composition, giving the idea of perfect homogeneity of constitution; but we soon find that we can track them through various changes, which commence by the appearance at their edges or borders of a white matter, which appears to gradually exude, and to which the greenish matter which before covered the entire surface soon looks disproportionate in amount, and collects itself as a more or less irregular mass in the centre.—*Vide* Photograph 69, Plate XII. The material which has exuded soon begins to granulate, and the granules are frequently set free in the Liquor Sanguinis. Occasionally we succeed in obtaining these corpuscles in the most incipient stages of transformation into the rosette form, and it will be observed that the corpuscles which undergo this metamorphosis are a little less coloured than those which we have spoken of as the diffused-edged corpuscles, some of which may be seen close by in the same photograph. —*Vide* Photograph 66, Plate XII., and description.

When a mass of granules undergoing this action lie together, the white matter which is exuded fuses, and forms a species of cement between them, so that the appearance is presented of dark granules, lying on or in a colourless substratum.—*Vide* Plate XII., Photograph 72. This cement matter may gradually disappear, then granules of a darkish colour alone remain to represent the so-called hæmatoblasts. —*Vide* lower part of Photograph 68, Plate XII.

So long as these rosette corpuscles, or their fragments, remain submerged in the Liquor Sanguinis, the white border is practically invisible, and the only part at all distinctly seen is the coloured central portion, which takes on angular and irregular shapes, owing to the changes which are occurring at the border. After a time these centres become further distorted and diminished by the passage from them of radiating processes. What is true of the corpuscles is also true of the granules produced by their disintegration.—*Vide* Photograph 86, Plate XIV. The central parts of these corpuscles, and of their granules, are the only portions which are visible so long as the serum is present, and they have been described by M. Hayem as " isolated, angular, colloid hæmatoblasts."

These visible central parts are very variable in size, owing to the fact that they sometimes represent the green lustrous portions of several corpuscles which have become fused together, at other times the green portion of single isolated corpuscles, and at others isolated fragments of single corpuscles.

The visible radiating fibres which arise from these bodies appear to have their origin, not from the exuded white matter, which might *a priori* have been anticipated, but from the central, slightly-coloured material, and this accounts for these fibres being visible in the Liquor Sanguinis, for if they had been

formed from the invisible corpuscles, or from the invisible part
or exudation border of the slightly-coloured ones, we could not
reasonably expect to see them. For a further verification of
the facts connected with the rosette corpuscles, and their
fragments or granules, the reader is referred to Photographs
66, 67, 68, 69, 72, Plate XII., and their descriptions.

Photograph 79, Plate XIII., shows how the rosette corpuscles and their granules adhere to the glass and become
surrounded by red corpuscles which enclose and support them,
while the white borders of the isolated ones are spreading down
and becoming lost to view, and simultaneously coloured fibres are
proceeding from the central darker parts. This action represents
the first step in the formation of radiating fibres from the
rosette corpuscles. In Photograph 80, Plate XIII. the action
has proceeded further, and the white borders of the rosettes
have almost entirely disappeared leaving the central dark
portions, which have become irregularly stellate and angular by
fibres being dragged out of them (by corpuscles, which first adhere
and then move on) to represent a group of so-called hæmatoblasts. In Photograph 81, Plate XIII., the coloured radiating
fibres are still better displayed, the white borders of the rosettes
are entirely dissipated, and the dark central portions are undergoing •disintegration and conversion into fibres. In this
specimen many red corpuscles are crossed by delicate fibres,
which are cutting into their substance, and as the contraction
of the fibre proceeds they will be completely cut into segments.

I referred in my previous paper to the converse case to the
one we have now been considering, in which instead of the white
matter accumulating at the exterior of the corpuscles, it
becomes gathered together into a single more or less globular
mass in the centre. The coloured material being arranged as a
zone or halo around it.—*Vide* Photograph 47, Plate IX.* It
will be interesting to trace what becomes of this zone of
coloured material, whether or not it spreads itself down, or

* First Paper on Development of Mammalian Blood.

becomes drawn out into fibres, as in the previous case. One thing these observations seem to render quite certain, namely, that like the lymph or gland corpuscles, these younger corpuscles of the blood are made up of two constituent elements, one of which is always darker or more coloured than the other, and this contrast increases *pro rata* with the assumption of colour by the young corpuscles, and we have afforded us, moreover, the most positive evidence that the material which has become coloured is located upon the exterior, while the colourless material normally occupies the interior, and thus its presence is entirely masked. It does not appear that these two substances are in chemical union, but that they are merely physically associated.

If I might venture upon a hypothesis, I would suggest that the substance (paraglobulin ?) of which the lymph corpuscle is composed becomes gradually converted from without, inwards, by a series of changes into hæmoglobin, the matter of the interior of the corpuscle being longer in undergoing this change. The corpuscles appear, however, to colour throughout their entire mass before they become *highly* coloured upon their surface, and the best fibre-forming corpuscles are those young ones which have become coloured throughout; for the colourless disc has less tendency to make fibre, and it is only the coloured part of the very young corpuscles and their granules which usually do this. For an example of the corpuscles most concerned in the formation of radiating fibres, *vide* Photograph 85, Plate XIV., and description.

The changes which occur in these young mammal corpuscles strongly remind one of Brücke's division of the matter of the oviparous corpuscle into Zooid and Oekoid.

Of these two substances the white matter seems to undergo subsequent changes by thinning and spreading down into very delicate, invisible films, and this occurs both with the corpuscles and their granules, and, as a consequence, the white matter disappears almost simultaneously with the formation of fibres from the coloured portion. It seems, therefore, certain that both these substances are concerned in fibrin formation, though they proceed to their goal in a different fashion. The two substances of which these corpuscles, or the granules which they yield, are composed, are not only distinct as far as appearance goes, but they have also a chemical distinction, which is shown by the difference of their behaviour with stains. The central part can be stained by aniline blue, while the border remains white. It is also quite easy to show that the substance which stains most readily is the one of which the radiating fibres are usually formed.

Sometimes entire corpuscles which appear to be coloured throughout their mass become wholly resolved into fibres ; at other times fibres are drawn off from the periphery only. In this latter case the volume and size of the corpuscle is, of course, diminished proportionately to the extent to which this has occurred. M. Hayem has given drawings of such partially exhausted corpuscles, and has described them as hæmatoblasts. For examples of these two kinds of action, *vide* Photographs 78 and 82, Plate XIII.

The perfect or unchanged primary corpuscle, having upon its surface only a comparatively thin layer of this delicate green substance, cannot stain very deeply, unless the dye is very intense. In all cases where the greenish exterior matter accumulates, or concentrates, the staining appears more decided, but this is, of course, only due to quantity or thickness.

Coagulation appears therefore to be preceded by a separation of the chemical constituents of the corpuscles, and of the granules which result from their disintegration. White fibrin being produced by a re-arrangement of the matter of the primary or colourless discs, and the coloured fibrin from the exterior of the intermediate corpuscles, and in addition to this there seems to be satisfactory evidence that some portion at least of the interior matter of the intermediate corpuscles also becomes resolved into fibrin.

I have before referred to the fact that certain of the partially-coloured corpuscles sometimes undergo changes somewhat different to that which we have expressed under the designation of the *rosette form*, instead of becoming flattened, more extended, and presenting a white granular exterior, with coloured centre, the converse effect to this appears to take place, the centre of the corpuscle exhibiting itself as a white mass, while the coloured matter of the corpuscle remains extended around it, as a sort of halo, and after a time, this circular band or halo may disappear, leaving nothing but the central white mass. In fact, apart from breaking up into granules, the younger corpuscles appear to be capable of two kinds of modification. In the first place the white or colourless matter may burst or protrude through the coloured pellicle, which then gathers itself up into a central mass or nodule; and in the second place the coloured pellicle may fall away from the colourless material, exposing it to view, and allowing it to gather itself up into a globule.

These colourless globular masses are of diverse sizes, according to the extent to which the colouring matter has penetrated towards the centre of the corpuscle, and as the development of the latter proceeds they become smaller and smaller, and are at length obliterated altogether.—Compare Photograph 47, Plate IX., and Photograph 73, Plate XIII. and their descriptions.

These white masses appear to have the power to aggregate closely with each other; whether the corpuscles aggregate first and subsequently losing their exterior allow the central masses to come closer together, or whether the aggregation takes place after the loss of the exterior colouring matter, I am at present unable to say.

It may be that the falling away of the exterior and the aggregation into masses takes place simultaneously. When however, the aggregation of these masses has occurred, it sometimes goes on to complete fusion, and the result is the formation of peculiar sheets which have been before observed by Osler of Montreal, and Max. Schultze.

All the corpuscles, I think, are competent to the production of these sheets, and they sometimes result from the coalescence of corpuscles of various shades of colour.

Most of the large white masses, seen in shed blood, probably originate by aggregation, in this manner.—*Vide* Photographs 73 and 74, Plate XIII., and their descriptions.

The bodies which contribute to the formation of the masses we have just described are much denser and larger in size than those which result from the breaking up of the white matter of the rosette corpuscles seen in Photograph 69, Plate XII.

It now remains to describe the second form, in which the younger corpuscles constantly present themselves during the process of disintegration, which in contradistinction to the *rosette*, I have designated the *granule-form*.

As in the case of the rosette form, all the corpuscles of the "Fugitive Group" are capable of undergoing this mode of change, but the primary corpuscle, nevertheless, is least of all liable to it, probably owing to its greater liquidity. The small groups of granules which have been supposed to be adhering hæmatoblasts, are in reality in many cases nothing more than corpuscles of this kind, in process of granulation. This I have ascertained by removing the cover-glass, both at ordinary and at ice-cold temperatures, and viewing them in the absence of the obscuring influence of the Liquor Sanguinis.

In many cases we are able to see corpuscles in various stages of disintegration, and can even count the number of granules into which they are about to become divided.

In Photograph 70, Plate XII., a corpuscle may be seen *in situ*, which shows ten or eleven distinct granules, and a second one which is more imperfectly granulated, considerable portions of the corpuscle still remaining intact. When the positions of the granules in relation to each other become disturbed, they still continue to adhere to each other by virtue of the matter exuded from their interior, and they frequently form themselves into circlets and beaded rolls, which have been described by M. Hayem under the name of wreaths, garlands, and chaplets.—*Vide* Photograph 71. Corpuscles of the kind which yield these granules may be seen unaffected on the same specimen.

In Photograph 71, Plate XII., the same conditions may be seen in corpuscles a little more coloured, while the granules and the granular masses depicted in Photographs 62, 64, and 65, Plate XII., are the products of the disintegration of primary or colourless discs, and in the latter case the granules are giving rise to delicate colourless fibres. Photographs 90, 92, and 93, Plate XIV., also show the direct breaking up of corpuscles into granules. In Photographs 92 and 93, the various stages of the action are displayed, while in 90, the granular masses, or so-called hæmatoblasts, are represented. Photograph 91, Plate XIV., shows another mode in which coloured granules are frequently produced, by the segmentation and division of entire corpuscles. These fragmentary parts of corpuscles are often bi-concave in form, and have then the appearance of small biconcave discs. In Photograph 87, Plate XIV., is seen a mass of granules derived from the breaking up of comparatively colourless discs. These are the so-called colourless hæmatoblasts. They often coalesce so completely as to form mere skins, or plates, which frequently take on irregular angular shapes as seen in Photograph 88, Plate XIV.

A further source of the so-called hæmatoblasts is found in the fact that corpuscles of all degrees of advancement are constantly being cut into fragments by threads of fibrin, which pass over and lie upon them; the fibre being fixed at its extremities cuts into the corpuscle when it undergoes its customary contraction. Such fragments can readily take on the bi-concave form. For examples of this process, *vide* Photographs 75, 76, and 81, Plate XIII.

All these granules and fragments yielded by the corpuscles of the "Fugitive Group" are capable of being extended into fibres, and this appears to be the final form in which they become hardened.

Photograph 75, Plate XIII., shows exceedingly well, young corpuscles breaking up into small granules, and the manner in which this granular *débris* becomes converted by extension into fine filaments of slightly coloured fibrin.

The cell wall and protoplasm of the ordinary white corpuscle also undergo granular disintegration, but the granules are easily distinguished from those yielded by the breaking up of the colourless discs, being smaller and more distinct, and never so far as I have been able to observe, entering into the formation of fibrin. Photograph 77, Plate XIII., represents the granules arising from the breaking up of the protoplasm of three white corpuscles.

That we have in the foregoing investigations been dealing with the same bodies as M. Hayem is rendered obvious by the presence of the white exudation or cement matter, and by the fact that both researches show these bodies to result in fibrinous products.

The changes which occur even in corpuscles having a considerable degree of colour are very imperfectly seen when the Liquor Sanguinis is present, and it is pretty certain that those which occur in the colourless discs, whether they be such as lead to the formation of rosettes, granule discs, granules, or of fibrinous networks, are entirely obscured.

All this is altered when the Liquor Sanguinis is removed, for we are then able to follow the changes which occur in the most primary corpuscles, that is, in the discs which are free from hæmoglobin.

We then learn that the most profound analogies exist between the behaviour of the colourless and the partially-coloured discs, which together constitute the "Fugitive Group;" that, in fact, they are capable of undergoing similar retrograde changes into rosettes, granule corpuscles, granules, and fibres, and also that masses of fused corpuscles undergo similar changes to single ones.—*Vide* Photographs 62, 63, 64, and 65, Plate XII., and descriptions.

As before stated it not unfrequently happens that the fragmentary portions of the corpuscles of the "Fugitive Group" are more or less bi-concave in form ; and this appears to have had considerable influence in causing M. Hayem to regard them as young red corpuscles. It has, however, not the least value in this connection, for the fragments of corpuscles will always take on this form, providing the Liquor Sanguinis is in the state favourable to its assumption and maintenance.

It will be well to bear in mind that these fragments have all the mutability possessed by entire corpuscles, and can, therefore, present themselves as spheres or as biconcave discs, &c. These forms are entirely dependent upon physical conditions, and have nothing whatever to do with vitality. The fact that they can be produced by artificial means warrants this inference.—*Vide* Page 23 and Photograph 40, Plate VIII. It is here seen that perfect biconcavity exists in bodies as large as the red corpuscles, and also in the smaller granules.

In addition to what has been said, it may also be stated that it is quite possible to break up experimentally the red biconcave discs by methods which preserve to the fragments their biconcave form.

The foregoing constitutes the principal proofs I have to offer of the mode and source of origin of the so-called hæmatoblasts, which it will be seen I hold to be entirely extra vascular and accidental formations, associated with the shedding of the blood.

Methods of Preventing the Formation of the so-called Hæmatoblasts.

I shall now ask the reader's attention to certain methods, which I have found either partially or entirely to prevent the formation of these bodies, by preserving intact the corpuscles of the " Fugitive Group."

The principal agencies which I have employed for this purpose have consisted of modifications of the temperature and of the use of preservatives, such as osmic acid and cane sugar.

Great stress is very properly laid by M. Hayem upon the importance of seeking for these bodies in blood which is perfectly fresh. A moment's reflection, however, will suffice to show us that neither blood which has become reduced to the ordinary temperature of the atmosphere, nor that which has been subjected to the temperature of zero, or lower, can be regarded in the light of fresh blood. Both these conditions, it must be admitted, are exceedingly abnormal when we reflect that the natural temperature of the blood is about $98\frac{1}{2}°$ Fah.

This illusion has arisen out of the circumstance of the well-known power of the freezing temperature to prevent coagulation and putrefaction; but it by no means follows that intense cold is competent to prevent all the changes in corpuscles of which they are susceptible; indeed, it is quite possible that it may induce or favour certain changes, and this really appears to be the case so far as granulation is concerned.

It is also a well-recognised fact that the normal state of the white corpuscle is best preserved by maintaining the blood at its customary temperature.

In perfect consistency with this fact, I find, too, that fewer so-called hæmatoblasts present themselves in the specimens when this method of preserving the corpuscles has been adopted.

There can be no doubt that cold holds in check those changes in the corpuscles which tend to resolve them or their granules into threads or fibres; but this is mainly, if not entirely, due to the fact that it temporarily suspends the properties of fluidity, adhesiveness, and viscidity, which so remarkably characterise the younger blood corpuscles, and in the absence of which fibrillation does not appear to be possible. Thus, although cold favours granulation, it may entirely hold in check the process of coagulation by preventing the formation of fibrinous networks.

When a maximum amount of solution of osmic acid is added to blood, the green granules are seen very distinctly; they are almost always single and isolated, and seem larger than usual. They may readily be observed in the interspaces between the red corpuscles, but are not as usual adherent to the glass, which may be known by the fact that they undergo constant Brunonian movements.

Many of the red corpuscles are massed together, and there are sound reasons for thinking that in this case the invisible corpuscles form the cement matter which binds them to one another. When the cover-glass is removed, the red discs are seen to be much contracted, and they gradually become darker coloured when exposed to the air.

Here and there may be seen small, smooth, green bodies, having a more circular outline than usual. These are the so-called hæmatoblasts.

As might be expected under such conditions, very few traces of fibrin can be observed.

From these observations I have felt justified in concluding that the maintenance of the normal temperature and the use of osmic acid are both means by which the number of the so-called hæmatoblasts can be greatly reduced ; but these agencies do not seem competent to entirely prevent their formation.

During a series of experiments, conducted for the purpose of ascertaining the best method of preserving the corpuscles of the "Fugitive Group," in order that they might be submitted to the action of stains,[*] it was found that nothing equalled ordinary white or loaf sugar, used in a certain manner.

This does not lie so much in its power to preserve the normal forms of the corpuscles as in the great influence which it exerts over their adhesiveness, not only with each other, but with the foreign bodies with which they may come in contact.

As this adhesiveness appears to be a necessary factor in the preliminary changes which the corpuscles undergo prior to the formation of fibrin, we can, by the use of varying quantities of sugar, so control these changes as to render it possible to track the whole series of corpuscular modifications and degenerations which eventuate in fibrin formation.

The method of adding the sugar to the blood is of the greatest moment, for in no case must water be introduced. The plan I adopt is to reduce the sugar to an exceedingly

---

Vide Section on Staining.

fine powder. A portion of this powder is then placed upon the end of the finger, and the prick is made through it, so that when the finger is squeezed the blood may come into immediate contact with the sugar and dissolve it. The liquid of the blood itself is thus made the solvent of the sugar. After thoroughly mixing with the point of the needle, the blood is allowed to flow, by capillarity, beneath the cover-glass in the ordinary way.

I shall now endeavour to give an idea of the influence of different proportions of sugar used in the manner indicated.

If we add a little only, say one part of sugar to three parts of blood, by volume, the adhesiveness of the red corpuscles, on which the formation of rouleaux depends, is not overcome, consequently these form nearly as usual. Certain white granules are also present, which are indicative of changes in some of the earlier forms of the corpuscles of the " Fugitive Group."

Little or no fibrin is, however, to be seen between the rolls. Delicate trail fibres may, it is true, be seen where the liquid has been driven back by an extension of the area of air bubbles. When the cover, however, is removed, we find that the primary corpuscles are laid down beneath the red ones, and on subsequently washing the slide, (after the method of Ranvier,) layers of fibrin are observed, freely perforated with small holes, which look like accumulating granules, also masses of granules, with white corpuscles adhering to them.

If, on the other hand, we mix together in the same manner equal portions of blood and sugar, we prevent the formation of rouleaux, and also the formation of fibrin, but we do not altogether prevent the occurrence of the granular disintegrations of the corpuscles of the " Fugitive Group."

Smooth, sheet-like bodies are formed, evidently by the fusion of corpuscles, of slightly different shades of colour. When these sheets are seen, after the removal of the cover and washing with water, they are decidedly fibrinous in character, their corpuscular origin being no longer apparent. In such specimens fibrin does not exist in any other form than in these sheets.

Of the so-called hæmatoblasts, those containing colour alone remain, the colourless ones being absent.

In the third case, much sugar being added to the blood, three parts of sugar to one of blood, the formation of rouleaux is entirely prevented. The colourless granulations are absent, and there is no formation of fibrin, for none can be seen, even after the removal of the cover, for no traces of either fibrin or corpuscles remain on the cover or slide.

In these preparations there is also an entire absence of the so-called hæmatoblasts, both coloured and uncoloured, and as, at the same time, all the corpuscles of the "Fugitive Group" are preserved, I feel justified in holding the view that the sugar has entirely prevented these accidental products being formed.

In the face of all this positive and negative evidence, we are inevitably impelled to the conclusion that the green lustrous granules and bodies which M. Hayem has termed hæmatoblasts are nothing more than accidental products derived from the breaking up of the corpuscles of the "Fugitive Group," *namely, of those which have obtained some colour*, but under certain circumstances it is clear that the primary discs also contribute a certain proportion of granules, but this can only be shown by removing the Liquor Sanguinis; hence none of the granules seen by M. Hayem were of this variety.

All the granules, coloured and uncoloured, are capable of being converted into fibres, and, therefore, take their share in fibrin formation, and in the coagulation of the blood.

It is a curious circumstance that it does not seem to have suggested itself to M. Hayem to enquire as to whether these minute and confessedly variable and irregular elements of the blood might not be due to the disintegration of some of the visible paler blood discs.

It appears to me unfortunate that an exhaustive investigation of this aspect of the subject should not have been made before venturing on a conclusion of such importance and magnitude as is contained in the statement, "These bodies are the germs of the red corpuscles, and I therefore propose

to call them hæmatoblasts." It is clear that their only claim
to be so regarded and designated must rest upon their
being proved to give rise to the red corpuscles, and this, of
course, involves the demonstration that, beyond all reasonable
doubt, they develop gradually and regularly into these bodies.
There are only two circumstances which can possibly justify
us in concluding that one body is the germ of another—either
we must observe, without break, the entire process of evolution
and growth, or we must be able to place the several elements
which we find in appropriate numbers in such serial order that
the transitions from one to the other shall be easy and obvious.
An attempt to do the latter, it is true, has been made by
M. Hayem, but it is to my mind very unsatisfactory.

In order that my readers may possess the opportunity of
forming their own judgment as to its value and conclusiveness,
I have introduced the diagram which M. Hayem furnishes
and on which he relies. It may be seen on Plate, XIV. Fig. 6,
and its description is as follows :—

Elements seen in the blood, prepared by the aid of
liquid A * (a) adult red globules, (b) young globules more or
less deformed, intermediate forms between the hæmatoblasts
and the adult red globules.

It will be observed, in the first place, that the specimens
brought together are selected from preparations which have been
subjected to the action of a re-agent, and secondly, that they
do not by any means fulfil the canon of easy and obvious transition which has just been laid down. It will be noted that they
are very irregular in form and size, and in state, as to corrugation or smoothness, and that in no case could we pass with any
feeling of security from one element to the next succeeding it
in order, and also that they seem to grow smaller where we might
naturally expect a graduated enlargement from first to last.

These elements are all represented as coloured, whereas
we know that many of the so-called hæmatoblasts, or granules

---

* Distilled Water 200, Sodium Chlor de 1, Soda Sulph 5, Mercuric Bichloride 0.50

seen in the blood, have scarcely any colour, even when an ice-cold temperature is employed, under which condition the Liquor Sanguinis often remains very colourless.

M. Hayem, being unacquainted with the fact that colourless discs existed in the blood, regarded the white granules which he saw as decolourised hæmatoblasts, but this we know not to be the case, for we can trace their production to the breaking up of discs which have certainly never possessed colour.

M. Hayem's view of the mode of the development of the blood labours under the disadvantage of being retrogressive in its character, and inconsistent with much valuable knowledge, which has been acquired in connection with this subject in the past by the laborious researches, both physiological and pathological, which have been made into the functions of the spleen and the lymphatics; it is moreover not in accordance with the analogies which he has himself made out to be true of oviparous blood, in which we see fully formed, *nucleated, colourless corpuscles,* gradually obtaining colour. All difficulty, however, disappears, when we come to know the true origin of the bodies to which M. Hayem has devoted so much attention, and recognise them as portions or fragments only of the younger discs of the blood. We are not then at all surprised to find that they possess different physical, physiological, and chemical characteristics to the fully developed red corpuscle, for this I have already shown to be the case with the discs of the "Fugitive Group," and what is true of these bodies as a whole (in these respects) must also be true of their parts.

I desire it to be distinctly understood that my observations on the work of M. Hayem are entirely restricted to the views he has published on mammalian blood, and have no reference whatever to his valuable research on the blood of the ovipara, which I regard as a most important contribution to our knowledge, and a powerful corroboration of the correctness of my own discoveries and views in relation to mammalian blood; for if the nucleated red corpuscle of the ovipara is preceded in the order of development by a nucleated colourless corpuscle, it is exceedingly probable that the non-nucleated red disc of the mammal will be preceded by a non-nucleated colourless disc. M. Hayem has shown the former to be the case in oviparous blood, and I have shown it be so in mammalian blood, but as both these corpuscles have an extra-vascular origin, it is not legitimate in either case to term them *hæmatoblasts.* They represent simply a stage in the development of the blood corpuscle, neither its beginning nor its end ; and in either case they are only entitled to be regarded as the colourless stage of the blood corpuscle ; the distinctive difference being that peculiar to each kind of blood, viz. : the one possesses a nucleus and the other does not. The reader may become acquainted with the relation which these two bodies bear to each other by a careful study and comparison of Photographs 61, Plate XII., and 83 and 84, Plate XIII.

While, then, I regard the description of these granular bodies, as given by M. Hayem, as accurate and reliable, I differ entirely from him in his fundamental conclusions as to the *rôle* they play in the development of mammalian blood. To make this difference distinct and clear, it may, perhaps, not be out of place to briefly recapitulate the view of the development of the blood, which I consider to be supported and justified by my own researches.

The reader may advantageously contrast Diagram 6, Plate XIV., given by M. Hayem, as illustrative of the mode of development of the blood, with the series of photographs of lymph and blood corpuscles, ranging in order from 94 to 100. It must be premised that according to my view the function of the blood-glands is not to produce a body like the ordinary white blood corpuscle ; but on the contrary a thickish disc-like body of about the same diameter as the blood-disc, which, in the course of its development, loses a delicate external layer, while its nuclear portion, which is very little less than the whole body was originally, undergoes conversion into a colourless bi-concave disc, and takes its place in the blood as its colourless disc. The lymph corpuscles, while in the glands may be seen to be under-

going these changes, as is shown in Photographs 94, 95, and 96.*
It will be readily seen that some of these bodies are already
beautifully bi-concave, while others are still in the process of
transition. On account of the necessity to get these structures
in the absolutely fresh state they have been obtained from
animals just killed, and are a little less than they would be if
got from the human species. In all other respects they compare
well with the colourless discs of human blood, seen in Photograph 97, and these in their turn form an easy transition to the
slightly tinted ones lying among the red ones, in Photographs
98 and 99, and the latter easily bridge the gap to Photograph
100, which represents the fully coloured-biconcave disc, the concavities of which have been slightly exaggerated by the osmic acid.
The red discs in 98 and 99 are also a little darker than they should
be, and somewhat altered from the normal owing to the use of
concentrated vapour of osmic acid to assist in the preservation
of the bi-concave form of the younger discs in the specimen.
It will be seen, therefore, that after protracted investigation of
the subject in its various bearings, I still hold to the view of
the development of the blood, set forth in my first paper, which
is briefly this—

1.—Suitable methods of examination will reveal in mammalian blood the existence of numerous corpuscles (about one
to sixty red ones,) which are incapable of being detected by the
use of the microscope alone, not because of their minuteness,
for they have the same diameter as the red discs, but because
they possess the same refractive index, and the same
colour as the Liquor Sanguinis, in which they are submerged.
When brought into view, and carefully examined, they are
found to be colourless bi-concave discs, and between these and
the red bi-concave discs the existence of other bi-concave discs,
possessing every gradation of tint, can be detected not only by
the eye, but more conclusively by the aid of the most delicate
photo-chemical tests.

---

* *Vide* the Author's Paper "On the Products of the Blood Glands."

2.—When the morphological elements of the blood-glands are examined, with the same precautions for their preservation as are found necessary with the blood corpuscles, they prove to be discs in various stages of development; the most advanced of which are excessively delicate, scarcely visible discs, of the same size and smoothness as the red blood discs, and of a bi-concave form.

3.—These bodies are poured into the blood, in large numbers at the subclavian and splenic veins, and are henceforth completely lost to view till they re-appear by gaining an amount of colour, slightly greater than that of the Liquor Sanguinis, when they become recognisable as the palest order of red corpuscles. During their colourless stage they are the invisible discs of the blood, and the analogue of the nucleated colourless oval corpuscles (so-called hæmatoblasts), of the ovipara, the nuclei of which alone are frequently the only parts that can be seen so long as they remain submerged in their own proper liquid. In both cases these corpuscles, as yet free from hæmoglobin, form the *missing link* between the advanced lymphatic and splenic and the red blood corpuscles.

SECTION V.

On the Morphological Products of the Blood Glands.

THE discovery in the blood of mammalia of large numbers of biconcave discs possessing the same colour and refractive index as the Liquor Sanguinis, and which therefore lie hidden in this liquid, together with corpuscles possessing every intermediate shade of tint between these and the red discs, has rendered it desirable (with a view of ascertaining definitely the source of these discs) to investigate thoroughly the nature of the morphological products yielded by the blood-glands, *e.g.*, the lymphatics, spleen, thymus, thyroid, supra-renal capsules, &c.

A limited devotion to the subject sufficed to show that this part of histology had been far from exhaustively studied, and that to this fact alone was largely attributable our profound ignorance concerning the mode of development of the blood, and the nature of its diseased conditions.

No advance worthy of the name has been made in this aspect of the subject since the days of Hewson, (1774.) Its literature is of the most meagre character, and the conclusions arrived at are conflicting and uncertain.

Reflecting this general ignorance and slavishly copying each other, our text books, with few exceptions to be presently noticed, concur in regarding the lymph corpuscles as " identical with the white corpuscles of the blood."

In the earlier part of this work (p. 25) I referred briefly to this question, and pointed out that Gulliver, Gray, and others did not concur in this view, and myself gave a number of reasons which appeared to me conclusive against it. It is but justice to the dead, and to the Scotch school of physiology, to state that

Professor Hughes Bennett entertained far more accurate views on this subject than many of his contemporaries. After speaking of the molecular basis of the chyle, described by Gulliver, he says: " Floating amongst these (oil globules) we observe granular and globular bodies about the 1-4000th of an inch in diameter, which on the addition of acetic acid exhibit a thicker margin than they did before. *In chyle taken from the thoracic duct there are also biconcave flattened discs, exactly resembling the coloured blood corpuscles in size and form, but destitute of colour.* Between these two kinds of corpuscles (the granular bodies about 1-4000th of an inch in diameter and the colourless biconcave discs) all kinds of intermediate stages may be observed, so that there can be little doubt that the former become flattened and are changed into the latter. *They are in fact embryo blood corpuscles which become coloured in the lungs.*"[*]

Again speaking of the lymphatic glands, he says: " On cutting into these glands shortly after digestion, and examining microscopically the fluid they contain, it may be seen that a molecular fluid (first described by Gulliver) is more or less crowded with *naked nuclei* which resist the action of acetic acid. On repeating the observation on fluid taken from the thoracic duct the same thing is noticeable, *only several of the nuclei are now flattened, and in every point, except colour, closely resemble the red blood corpuscles.*"[†]

It is quite obvious from these passages that Professor Bennett did not, as so many have done, confound the true lymph corpuscles (which he regarded in their most developed form as colourless biconcave nuclei) with the ordinary white corpuscles of the blood with which, in his investigations of Leukhæmia, he had become so familiar.

In the thoracic duct, side by side with these *colourless biconcave disc-shaped nuclei,* destined to become coloured after they had entered the blood, Bennett recognised also the presence

---

[*] Text Book of Physiology, page 60.
[†] Text Book of Physiology, page 208.

of bodies which he regarded as the equivalents of the ordinary white corpuscle.

We shall see in the sequel how closely these views of Bennett approximate to the truth.

The illustrious German pathologist, Professor Virchow, describes lymph corpuscles taken from the interior of the follicles of the lymphatic glands as consisting of two kinds. 1st. Free nuclei with and without nucleoli. 2nd. Cells with smaller and larger nuclei which are closely invested by the cell-wall. In common with other investigators, he does not find in these glands any bodies corresponding to the multinuclear white blood corpuscles, and I am not aware that he has referred to the condition of these elements as they exist at the upper part of the thoracic duct, but, inasmuch as he states "that of the minute elements contained in the follicles some appear to become separated and afterwards to mingle with the blood as colourless blood or lymph corpuscles," it is apparent that he considers *some* of these elements develop into the white corpuscle of the blood ; on the other hand, Virchow offers no opinion as to the origin of the red corpuscles, but, on the contrary, says, "the whole history of the red corpuscles is still invested with a mysterious obscurity, inasmuch as no positive information has, even at the present time, been obtained with regard to the origin of these elements."[*]

The following quotation shows, however, that he leans towards the view first promulgated by Hewson, viz.,[†] that they are in some way or other the products of the lymphatic glands and spleen. " A good many years elapsed (after 1845) during which I found myself pretty nearly alone in my views. It has only been by degrees, and indeed, as I am sorry to be obliged to confess, in consequence rather of physiological than pathological considerations, that people have come round to these ideas of mine, and only gradually have their minds proved accessible to the notion that in the ordinary course of things the lymphatic

---

[*] Virchow's Cellular Pathology, p. 223.   [†] p. 172 ibid.

glands and spleen are really immediately concerned in the production of the formed elements of the blood ; and that in particular the corpuscular constituents of this fluid are really
descendants of the cellular bodies of the lymphatic glands and
spleen which have been set free in their interior, and conveyed
into the current of the blood."

If from these writers, so justly entitled to a primary hearing
on these questions, we turn to later investigators, we do not find the
area of our knowledge to be materially enlarged; for, while much
investigation has taken place in connection with the white blood
corpuscle, it relates chiefly to its anatomical constitution, to
those physiological attributes known as amœboid motion, and to
the facts connected with its passage through the vessel walls.
Nothing further has been done to indicate the connection which
the products of the blood glands bear to the red and white
corpuscles of the blood.

At the very outset of this research, and in the face of the
reservations of the eminent observers I have quoted, we are
met in all directions with the stereotyped statement that " *the
products of the blood-glands are identical with the white corpuscles
of the blood*," a statement at once incorrect and confusing,
because it leads to the erroneous opinion that the function of
the blood-glands is simply to originate the white corpuscles of
the blood from which (to borrow the words of Prof. Huxley) the
red corpuscles are " in some way or other " produced.*

In the first place, the white blood corpuscles are not even
identical with each other, either in size, in chemical constitution,
in histological character, or even in physical appearance.   They
so differ in size that, while some are not more than 5·20000th of
an inch, others attain to a diameter of 10·20000th of an inch.
Some can be made to exhibit an uni-nuclear appearance only,
while others may display from two to five nuclei. ·Both of these
varieties may lie in weak aniline blue staining fluid for hours†

---

* Elementary Physiology, page 62.
† In one test case for thirty hours.

without their nuclei becoming stained; but if their pellicular cell covering be ruptured mechanically or otherwise, which can readily be done at will, the freed or exposed nuclei (as the case may be) at once become deeply stained. Other forms appear on the contrary to be already in the condition of free nuclei, for they take the stain throughout their substance the moment it is added to the blood. They are obviously devoid of the protective covering which the former class possess.

The utmost that can be with accuracy affirmed is, that some of the white blood corpuscles have a close resemblance to, and an affinity with the true lymph corpuscle, *i.e.*, with the body which it is the special function of the blood-glands to produce, while others, the multi-nuclear variety, bear a resemblance to certain bodies which are not (in the state at least in which they now exist) the normal products of the blood-glands, but yet may be found in the thoracic duct and the larger lymphatics. In a word, if we could abstract from the blood its red corpuscles, there would still remain in it colourless discs and uni-nuclear and multi-nuclear white corpuscles. The prototypes of the two former can be found in the blood-glands, but of the latter only in the thoracic duct and the larger lymphatics. The true relations however which subsist among these bodies will become more and more obvious as we proceed.

When we cut into the substance of a fresh lymphatic gland or of the spleen, and take out on the point of a knife a portion of the lymph which flows spontaneously into the bottom of the incision, we find that we have present two kinds or conditions of corpuscles, one of which is readily seen, being distinct and corrugated, the other smoother, fainter, and far less distinctly visible. The former kind have a diameter of from 4 to 5·20000th of an inch, and are the bodies in the lymph which have usually been described as free nuclei. The latter, if observed at all, which is doubtful, have not hitherto attracted any attention, probably owing to the great variety of appearances they present, which is most perplexing till the key to their true nature has been discovered.

At this stage of the inquiry I must ask permission to anticipate somewhat the proof which I have to bring forward, so far as designating these two kinds of corpuscles by the terms primary and advanced, as it will enable me to present the facts in a clearer and more intelligible form, It is my intention to show that these two kinds of corpuscles represent extreme conditions in the development of one and the same body, and that the same terms permanent and fugitive are as applicable to them as to the corpuscles of the blood, but with the difference that in this case the youngest or most primary is the most stable, whilst the more advanced ones constitute the fugitive class. It will be seen that this is precisely the converse of the blood.

Again, between these extremes of the lymph corpuscle as with the blood it is possible to detect every intermediate kind, and we are consequently compelled to conclude that we have also in these corpuscles to deal with bodies which are undergoing a regular and gradual development, which, as it proceeds, renders them more plastic, smoother, whiter, less visible and more amenable to change when the lymph is shed. Here in fact as in the blood we are presented not with one body but with a graduated series, the least perfect of which are the bodies best seen and commonly described as the lymph corpuscle, while the more advanced ones approximate with exceeding closeness to the invisible discs of the blood, in size, form, appearance, physical and chemical constitution, and more particularly in that amenability to change from their usual corpuscular condition into the forms and conditions commonly recognized as fibrin.*

As a consequence of this extreme instability, the advanced lymph corpuscles are rarely seen in shed lymph in their true and normal form, but generally in every conceivable state of degeneration and breakdown, and it is only after we have laid hold of this fact and have adopted means for their preservation that we can form any conception of their real nature.

---

* *Vide* Section on Fibrin Formation.

In Photograph 101, Plate XV., the reader will find a representation of the body found in the follicles of the glands, which is commonly regarded as the lymph corpuscle, a d usually conceived to be a simple nucleus. It is, however, as I have said elsewhere, to be regarded as a corrugated, thick, disc-shaped body, having an average diameter of from 4 to 5·20000th of an inch. This is the body which I propose to distinguish by the name of the *primary lymph corpuscle*. It has comparatively little liability to change, is rather rigid than plastic, and while submerged in the liquor lymphæ has very little tendency to flatten and spread itself down upon the glass slide. It neither stains or swells under the influence of saline aniline blue*. Photograph 102, Plate XV., shows such corpuscles obtained from a fresh lymphatic gland of a pig. It will be seen that the puckerings or corruga-tions have disappeared, and that the corpuscles have at the same time increased in diameter, and are, therefore, more disc-shaped. Some of them are of much greater diameter than others. These latter are the more advanced ones, they have attached themselves to the glass, and are spreading down upon it.

In Photograph 103, plate XV., we have an illustration of the appearances presented by lymph corpuscles taken from the same gland after the liquor lymphæ has been withdrawn from them by the "method of isolation" as practised with the blood.† The difference between the two conditions of the corpuscle is here made more manifest. The primary ones photograph with a strong dark edge, and have a tolerably regular outline, and are all of about the same size, while the advanced ones are without a distinct outline, and are modified in shape by being subjected to pressure by the more rigid primary corpuscles. It will be seen further that when they are unsupported by these they have a marked tendency to spread themselves out upon the slide, to become more irregular in outline, and more confused

---

* Half per cent. aniline in ¾ per cent. salt solution.
† Vide page 7, Section I.

and granular in character. Between these corpuscles there is also a great difference in colour, the advanced ones being white or colourless, while the primary ones have a greenish tint, which, like yellow or red tints, photographs of a dark colour. This colouring matter, whatever it may turn out to be, must not be confounded with hæmoglobin, as it does not yield the absorption bands of this substance.

In Photograph 104, Plate XV., we have a good example of the action of spreading or extension, to which the more advanced lymph corpuscles are liable. They appear first to adhere to, and then to spread themselves out upon the glass slide. Any granules that may be present are driven outwards by the spreading action, and collect mainly at the circumference of the extended corpuscles, hence they are often seen to be disposed in the form of arcs, or portions of large circles. These curves map out and indicate the degree of extension which the corpuscles have undergone. Photograph 105, Plate XV., shows an excessive action of the kind which came on rapidly. The primary corpuscles, too, in this case, have spread out into larger discs, while the advanced ones are spread enormously, as indicated by the granular boundaries. In Photograph 106, Plate XV., we have an illustration of this peculiar action in its first stages. None of the above specimens are interfered with by reagents. In Photograph 107, Plate XV., is shown a specimen of lymph which was diluted with fresh blister fluid. The advanced corpuscles have spread excessively, and the primary ones have been converted into smooth discs of about the same size as the red blood discs. Photograph 108, Plate XV., illustrates a case in which this spreading action, peculiar to the advanced lymph corpuscle, has been restrained by diluting with a five per cent. solution of salt. The corpuscular character of the advanced corpuscle is now better seen, yet the difference between it and the primary one is still sufficiently obvious. Photograph 109, Plate XV., is a specimen of gland corpuscles from the pig, prepared by diluting the lymph with an equal quantity of $\frac{3}{4}$ per cent. of salt solution,

and then preserving and diffusing simultaneously by blowing
hard upon it with an indiarubber syringe, till quite dry. The
differences between the primary and advanced lymph corpuscles
are not as well seen in this specimen as in 108, but it will be
observed that the ground work is covered with granules, some
of which are dark, and others white, which are probably the
débris of both kinds of corpuscles.

Photograph 110, Plate XV., also shows the existence of
these extreme kinds of corpuscles in lymph obtained from the
thyroid gland of the calf. The corpuscles which are intermediate
can also be well seen in this example, and are detectable by
their varying degree of colour and tendency to spread.

By adding cane sugar in a pulverised state to the lymph I
have been able to preserve very considerably the normal
characteristics of the corpuscles, and the advanced ones are rendered much less liable to spread and enlarge, consequently a
greater uniformity of size between these and the primary
ones is preserved. Photograph 111, Plate XV., is also
a specimen from the thyroid gland of the calf preserved in this
manner. The preparation was spread out in a thin layer upon
the glass slide by forcible blowing with the indiarubber syringe,
a cover glass was then laid over it and $\frac{3}{4}$ per cent. solution of
salt run between. Corpuscles of all shades of colour from very
dark primary to colourless ones may be observed. It seems
scarcely credible that the corpuscles seen in 110 and 111 should
be derived from the same source.

In the foregoing remarks I have endeavoured to show that
great differences exist among the lymph corpuscles when taken
quite fresh from the blood glands of recently killed animals—
differences which have their origin not so much in variation of
size, as in that of colour and in power to resist the influence of new
or altered surroundings. Indeed it would appear that these bodies
become, in proportion to the degree of their development,
excessively amenable to change when the lymph is shed or
removed from its normal habitat. Now it might naturally be
anticipated that with these physical changes of constitution

which render the corpuscles more delicate, fragile, fugitive, and less visible, there would be concurrent chemical modifications on which these would be based, and this is really the case, for we find that the relations which these bodies bear to stains is entirely in keeping with their physical deportment.

In Photograph 112, Plate XV., is a specimen of lymph from a lymphatic gland which has been diluted with blister fluid, and subsequently stained with saline aniline blue. The aniline blue causes the more advanced corpuscles to become greatly enlarged and flocculent or granular in appearance, and simultaneously they become stained of a deep blue tint. This action of simultaneous staining and enlargement takes place rapidly, and is proportionate to the amount of aniline present. A minute amount of aniline, however, which will scarcely produce visible staining is capable of causing great enlargement.

It is curious that when so enlarged they are not globules as might *a priori* be expected, but greatly extended discs, and simultaneously with this action in the advanced corpuscles, the most primary ones, which refuse the stain, undergo shrinking and contraction. By this means a broad and well marked distinction is established between the two classes of corpuscles, the primary ones not only refusing to stain, but also undergoing shrinking and becoming distorted and irregular, while the advanced ones become at once stained and swell up into smooth, regular discs of large diameter, which frequently lie in part over each other as coins might do, thus proving their disc like character. The degree of swelling or extension may be seen by comparing Photographs 101 and 112.

The proportion of these corpuscles in various glands appears to differ sometimes in favour of the primary, and at others of the advanced ones.

In Photograph 113, Plate XVI., is seen a specimen of fresh sheep's lymph from a lymphatic gland. This specimen was first diluted with a ¾ per cent. solution of salt, and then run under a cover glass. On examination the corpuscles were found to present their usual appearance. Being subsequently stained with saline aniline blue, many previously barely visible corpuscles rapidly put in an appearance, and these became greatly swollen and deeply stained. Between the most decided primary ones which are dark, irregular, contracted, and unstained, and are of the smallest diameter, and the advanced ones which possess the greatest diameter, every intermediate diameter and appearance in relation to swelling and staining may be observed. Adhering to the surface of the advanced corpuscles are numerous granules which are obviously similar to those which are free in the liquor lymphæ of such specimens, and the source of which will be referred to later on. In specimens of lymph in which these granules are not present the corpuscles are smooth and clean, and this is an additional reason why they are to be regarded as merely accidental adhesions. If this view is correct it implies that the advanced lymph corpuscles become, under these abnormal conditions, adhesive and pick up the granules as they float among them. It often happens in specimens of lymph that the advanced corpuscles are only obvious by means of these adhering granules; in other words, the corpuscles themselves are too delicate to be seen.

As a further illustration of the effects of saline aniline staining upon the advanced lymph corpuscles Photographs 114 and 115, Plate XVI, may be compared. The former shows advanced lymph corpuscles (and also granules, which appear to have resulted from the disintegration of others,) which have been separated from the primary ones by the method of isolation in virtue of their power to adhere to the glass. Their size and form have been more or less preserved by means of the ice-cold temperature.

In Photograph 115, Plate XVI., the *same* corpuscles and granules are again shown, after being subjected to the action of the aniline stain. It will be observed that both the corpuscles and the granules have become greatly swollen, and simultaneously with this have become stained of a deep blue tint. This shows absolutely that the corpuscles and granules which stain are also those which swell and enlarge, and places in our hands a means of distinguishing from each other not only the corpuscles but also the granules, which result from their disintegration. This differentiating action of the saline aniline blue is very well seen in Photograph 116, Plate XVI., in which the most advanced corpuscles are those which have undergone the greatest degree of extension and swelling, and between these and the dark primary ones, which have become contracted, corpuscles of varying diameters, may be seen which represent intermediate stages of development.

In Photograph 117, Plate XVI., corpuscles from the lymphatic gland of a pig have been submitted to the action of the saline aniline blue, and we are able to distinguish three states of the corpuscles. 1. The most primary, which have neither spread themselves down upon the slide nor have become stained. 2. An intermediate kind, which have to some extent spread themselves down, but have not absorbed the dye. 3. The more advanced ones, which in the Photograph show a granular appearance, and which have all taken the stain, but in various degrees ; those whose physical properties are the furthest removed from the primary ones having taken the stain most greedily.

The illustrations and descriptions given will suffice to indicate the nature of the general conclusion arrived at by an extended study of these bodies—viz., *that they are structures which are undergoing a gradual and continuous process of development which is associated with changes in both their physical and chemical characters.*

The principal difficulty that we have in the study of the lymph corpuscle arises from its being almost impossible after the lymph is shed to preserve in their full integrity its most developed forms. These can scarcely be brought into contact with ordinary surfaces such as those of glass without being resolved into some one or other of the fibrin forms ; indeed, as before said, they are the source of fibrin in lymph as the colourless disc is in blood. It is owing to this that their existence as corpuscular structures has been overlooked, for the fibrin when seen has been accounted for in a very different way. When, however, we become aware that the fibrin present in any specimen of lymph really represents the changed and modified forms and states of its most advanced corpuscles, we at once see that if we would ascertain the relation which these bodies bear to the colourless discs of the blood, we must, if possible, devise means by which they can be preserved for examination in as nearly as possible their normal state.

Considerations of this kind enable us to understand fully the difficulty which has attached in the past to the comprehension of the mode of development of the blood, for whether we examined the blood on the one hand, or the lymph on the other, the connecting links were in both cases largely obliterated, owing to the corpuscles which bridge over the chasm being in both cases resolved into fibrin, and therefore no longer recognisable as definite structures, but appearing as a third substance of unknown and problematic origin.

Under the generic term "lymph corpuscles" a number of bodies have been grouped together and regarded as *identical* some of which present very marked and characteristic differences, and it has become a common error to regard the white blood corpuscle as the type of all these bodies. This practice has given rise to the utmost confusion of thought, and has rendered it impossible to form even an approximately correct view of the nature of the development of the blood. The following bodies have thus been confounded :—(*a.*) Corpuscles from the lymphatic glands; (*b*) Splenic corpuscles; (*c*) Thymus corpuscles; (*d*) Thyroid corpuscles; (*e*) Corpuscles of the supra-renal corpuscles; (*f*) White blood corpuscles; (*g*) Leukhæmic corpuscles; (*h*) Pus corpuscles. Further confusion has also arisen from failing to recognize the fact that these bodies exist in all stages of development.

Of the various kinds of corpuscles enumerated a, b, c, d, and e, are alike, and are readily divisible into the two kinds, representing extremes, which we have designated the primary and advanced. It is the destiny of all these bodies to become red blood discs after having passed through the stage of the colourless disc ; f, g, and h, on the other hand represent certain deviations or departures from this nominal mode of development, represent, in fact, the accidents (so to speak) of the great blood making process. These relations will, however, be better understood as we proceed.

From what has already been said it will be seen that the lymph corpuscles in the spontaneous changes which they undergo when the lymph is shed, divide themselves broadly into two kinds, and the existence of these two varieties or stages of development is further corroborated by the use of certain staining fluids. If we proceed to inquire more critically into the constitution of these corpuscles, we find that their difference of physical and chemical behaviour appears to rest upon histological diversity ; that, in fact, we have in the primary lymph corpuscle, which has hitherto generally been considered as a nucleus to deal with, a true cell, the outer wall and protoplasm of which is in such close contact with that of its nucleus as to cause them to appear to be, and even to act under some circumstances, as if they were one only. The existence of the external cell-wall can, however, be made very obvious in several ways. In the first place corpuscles possessing it, photograph of a much darker tint, almost, and in some instances, quite as dark as red corpuscles. What the colouring matter is which causes this photogenic reaction we do not know, all we can say at present is that it does not yield the spectrum of hæmoglobin. It is, however, nearly as non-actinic as this substance. Examples of these photogenically dark primary lymph corpuscles may be seen in Photographs 103, 110, and 111, Plate XV. The colouring matter on which this non-actinic property depends is not very perceptible to the eye, and is entirely removed when lymph corpuscles are allowed to remain for some time in contact with Liquor Sanguinis. They then photograph as free from colour as the white blood corpuscles. I have not

found this coloured condition of the primary lymph corpuscle
to be present in the spleen, probably owing to the free blood
supply of this organ. A further feature which distinguishes
the primary lymph corpuscle is the fact that it may, without
alteration, remain a long time, *e.g.*, thirty hours in a staining
fluid, which at once causes the advanced lymph corpuscles to
become deeply coloured, and to swell up into discs of large
diameter. Not only do the primary corpuscles not stain or
swell, but they actually contract under these conditions,
vide Photographs 113, 116, and 118, Plate XVI. If, however,
while lying in the staining fluid we mechanically compress these
corpuscles, the external membrane may be seen to rupture, and
the contained nucleus at once stains and swells up precisely like
the advanced corpuscle. The only difference now apparent
between the two being that some granules which have refused
the stain may still remain attached to the former ; these represent
the granular débris of the protective and unstainable capsule.
Generally these granules remain attached to the edges of the
corpuscles, but occasionally they adhere all over them, as
shown in Photograph 118, Plate XVI.

The *Primary Lymph Corpuscle* appears, therefore, to consist
of three distinct parts. 1.—An exterior layer, wall, or pellicle,
which will not stain with the aniline blue nor allow the staining
fluid to pass through it so as to affect the interior (nucleus).
2.—An interior mass, (nucleus,) which when exposed by rupture or
removal of the capsule, stains easily and deeply, and swells up
greatly under the influence of the saline aniline. 3.—A clear
colourless liquid, which exudes from this stained mass either
from its interior or by a process of separation and aggregation.

This refractile exudation can take on similar forms to
those exhibited by myelin.

The *Advanced Lymph Corpuscle*, on the other hand, appears
to consist of two parts only, viz., that which stains and which
appears to be situated externally, and that which exudes as
clear hyaline liquid, and behaves like myelin, in other words, in
the advanced corpuscles the external wall or pellicle has
disappeared.

Now, if the *Primary Lymph Corpuscles* are to be regarded as true cells, or bodies in possession of a cell wall, of a nucleus, and of intra-nuclear contents, then we must consider that the cell wall is ordinarily in close contact with the nucleus, and that the development which the corpuscles undergo, when all proceeds normally, consists in the gradual destruction and removal of this outer capsule, and in this case the advanced gland or lymph corpuscle will have to be regarded as a freed and developed nucleus, *in a word, the primary lymph corpuscle is a cell containing a nucleus, and the advanced lymph corpuscle is this nucleus set free at a certain stage of its development.*

The following definitions of these bodies appear to be possible ---*Primary lymph or gland corpuscle.* A thickish non-nucleated corrugated disc of an average diameter of 4 to 5·20,000th of an inch, which generally possesses some colouring matter causing it to photograph of a rather dark tint,* but which bleaches if brought into contact with Liquor Sanguinis. This corpuscle refuses to become stained or to *swell* up into a smooth disc under the influence of saline aniline blue, but on the contrary contracts, and becomes irregular and more corrugated ; when submerged in the liquor lymphæ it appears corrugated and rigid ; but when it passes from this influence into a bubble space, in other words, into air loaded with moisture, it immediately becomes lustrous and smooth, exhibits a greenish tint, which photographs dark, and spreads down upon the slide with great increase of diameter, *vide* Photograph 119, Plate XVI. When kept under sealed cover glasses in its own liquid (liquor lymphæ) it becomes whiter, thinner, smoother, and tends to biconcavity. It rarely, if ever, exhibits amœboid motion—never, under any circumstances, displays a multi-nuclear appearance, but is susceptible of the peculiar change shown in Photographs 122, 123, and 124, Plate XVI., the

---

* Examples of these dark primary lymph corpuscles may be seen on Photographs 103, 108, 110, 111, &c., Plate XV, and 116, 120, and 121, Plate XVI.

nature of which is discussed on page 117.   When introduced
into blood these corpuscles become the same colour as the
Liquor Sanguinis.

Advanced lymph or gland corpuscle. A very delicate, smooth,
non-nucleated, fugitive disc, of the same diameter as the red disc.
Excessively plastic, falling readily into fibrin forms, stains
with extreme ease with the saline aniline blue, and swells up
enormously—say to a diameter of 8 or 9·20000th of an
inch, but still retaining its discoid form.   Much more liquid
and mutable than the primary disc, probably possesses
amœboid motion, but this cannot well be made out, for it
is too indistinct in the liquor lymphæ to be satisfactorily
watched—easily spreads down, and is lost as a film upon
the slide, *vide* Photographs 105 and 107, Plate XV.   When
introduced into blood, or pure Liquor Sanguinis, it is no longer
seen.   It is this body (which represents the developed product
of the blood glands) which passes on into the blood to become
in the first place its *colourless disc*, and ultimately, after
acquiring colour, its red corpuscle.*

Anyone who has the most superficial acquaintance with
the products of the blood glands will be struck with the
protean forms and appearances under which these bodies
present themselves, due in part to their extreme plasticity,
and in part to their great dependence upon the state, for the
time being, of the liquids in which they are submerged,
which, as in the case of the blood, determines not only their
size, but also their state as to smoothness or corrugation.
The absence also of elasticity in the lymph corpuscle renders
it far more amenable to permanent modification of form
than the advanced blood corpuscle, but the same is true of
the earlier forms of the latter body, which approximate in
origin and nature to the lymph corpuscle.

It has been before explained that when saline aniline blue

---

* *Vide* Section "On the identification of the advanced lymph disc
with the colourless disc of the blood."

is used as a means of distinguishing the various developmental stages of the gland corpuscles, it is found that the least advanced or more primary corpuscle is the one which refuses to stain, and which contracts under the influence of the staining fluid, while the most advanced is the one which not only stains readily, but swelling up becomes distended into discs of large diameter. If for the saline aniline blue we substitute a simple aqueous solution of aniline blue, we are able to bring out a still more marked difference in these corpuscles which is of the utmost importance inasmuch as it furnishes us with an additional demonstration of the cellular character of the primary lymph corpuscle, and shows the relation which this primary cell bears on the one hand to the uni-nuclear variety of the white blood corpuscle, and on the other to its own nucleus, which in the ordinary and normal course of events is to be set free and to become the colourless disc of the blood. Unlike the saline, the aqueous staining fluid favours endosmotic action, and as a consequence the delicate wall of the cell, which heretofore has been in very close proximity to the nucleus, becomes separated from it and undergoes distension, sometimes to a very marked degree, so that the corpuscle now exhibits a clear hyaline border, in fact, presents itself as a veritable cell, nucleus, contents, and wall. *Vide* Photographs 122, 123, 124, Plate XVI. Photograph 122 shows the primary lymph corpuscle of the spleen of the pig ; 123 of the lymphatic glands of the pig ; and 124 of the thyroid gland of the calf. *None of the corpuscles which take on this cellular appearance become stained with the aniline blue.* The main variations which they exhibit relate to the extent of the area of the clear zone which surrounds the nucleus, and to the state of the latter body itself which is sometimes smooth, at others corrugated, and in some cases, where the distension of the cell-wall is very great, it appears even to be undergoing disintegration. Under ordinary conditions the nucleus fills up the cell and entirely determines the appearance it will present as to colour, smoothness, corrugation, &c. ; being, in fact, the body which we see through the delicate cell-envelope, and ordinarily regard as

the corpuscle ; consequently the irregular and furrowed appearances presented by the unaltered corpuscles are still retained by
the nuclei when the cell-well is separated from it and distended
into an exquisitely delicate membrane. The corpuscles, which
under the action of osmosis yield the largest zones, are probably
the cells with the thinnest walls and those which will be the
first to liberate their nuclei.

All the free nuclei stain and swell, but in varying degrees ;
the most recently liberated swell least, and therefore appear to
be more strongly stained. On the other hand the most advanced,
or those which have been set free for a longer period, swell
greatly and look less stained.   They are all, however, probably
stained equally, the difference of appearance in this respect
being due to the degree of swelling.

Many reagents favour these osmotic modifications of the
corpuscles.   Photograph 122, Plate XVI., illustrates the effects
produced by an aqueous solution of aniline blue.   The advanced
corpuscles, which in the original specimens are deeply stained
of a blue tint, and therefore look very conspicuous and contrastive, lie between and among the corpuscles, the cell-walls of
which have been made to show themselves.   In the Photograph
the contrastive effect given by the colour is of course absent.
Photograph 123, Plate XVI., is a specimen of gland corpuscles,
in which endosmose has been excited in the primary corpuscle
by means of gum and liquor potassæ.   Photograph 124, Plate
XVI., is a specimen in which a small portion of the contents of
a lymphatic gland of a calf was allowed to run under a cove
glass and fill up a portion of the area only, the other portion
being filled with a two per cent. solution of osmic acid.
The region in which the endosmotic action occurred was
that in contact with the osmic acid solution.   This reaction
does not invariably succeed.   During the process of normal
development in the glands, the cell-wall of the primary corpuscle
probably undergoes gradual thinning and solution, for when the
glands are absolutely fresh very little granular débris exists in
the liquor lymphæ ; but if the same glands are kept for a time

a great deal of granular débris is found, most of which appears to be due to the disintegration of the cellular envelopes. This of course sets the nuclei free, and under these circumstances the staining is more general and uniform. It is worthy of note that no wall or pellicle can be made to show itself upon the corpuscles which take the stain, *i.e.*, upon the nuclei.

SECTION VI.

On the Origin, Development, and Destiny of the White Blood Corpuscle.

The bodies we have to consider in this connection, and to compare with each other, are the primary and the advanced lymph corpuscles and the nucleated white blood corpuscles. The latter bodies are divisible into two varieties according as to whether they possess one nucleus only or several (2 to 5), but as no difference has been detected in their physical or chemical behaviour, they may justly be regarded as the same bodies in various stages of development.

Photograph 125, Plate XVII, shows the usual spherical appearance of the white corpuscles in the absence of amœboid movement ; while Photograph 128 gives examples of its uni-nuclear variety in which the nucleus has been revealed by staining. In Photograph 129 the transition stages are shown by which the uni-nuclear becomes converted into the multi-nuclear kinds seen in Photographs 130, 131, and 132. In Photographs 128, 129, and 130, the nuclei are displayed by staining with phosphene. In Photograph 131 they are displayed by the "method of isolation," and in Photograph 132 they have been stained with aniline blue, which, in keeping with its action on nuclei generally, has caused them to become greatly swollen, and to encroach upon the protoplasm, which is partly dissipated and partly still adherent to the nuclei. In Photograph 136 the granular protoplasm has been stained with aniline scarlet, and therefore photographs of a dark tint. It is noticeable that the nucleus of the uni-nuclear variety is generally very large, that, in fact, the nuclei of the multi-nuclear kind taken together would but form a mass of much the same size, and that speaking generally the greater the number of nuclei present in a corpuscle the smaller they individually appear to be. Nothing, therefore, seems more likely than that the uni-nuclear should develop into the multi-nuclear kind.

As to the protoplasm of the white corpuscle, it appears to be capable of existing in several distinct states, (1) as a liquid ; (2) as granules ; (3) as fibres. When in the liquid state it may be ejected from the interior of the corpuscle by means of saturated solution of salt. The first action of this re-agent is to cause the corpuscle to visibly distend ; it may then be seen to burst and emit colourless liquid content matter from some point of its circumference. This generally continues to adhere to the shrunken capsule and the enclosed nuclei, but in some cases it gets free and swims about as a delicate homogeneous globule. Photograph 135, Plate XVII., illustrates this reaction. On the other hand liquids of this character are exceedingly prone to fall into a granular state; hence the protoplasm of the white corpuscle is frequently seen in this form, as in Photographs 128, 132, and 136, Plate XVII. The third or fibrous condition has given rise to the erroneous view that protoplasm possesses considerable histological complexity, being a delicate structure formed of an intra-cellular network of very fine fibrils (Heitzmann's network). Whenever this appearance is present it is probably due to initial coagulative changes, the granules become adhesive, and, being viscous, their mutual motions quickly extend them into delicate fibres and networks. The same action not unfrequently takes place with the content matter of the nuclei.

If we seek further back for a body which may be regarded as the antecedent of the uni-nuclear white blood corpuscle, we find it in the *primary* lymph cell. This body does not differ from the uni-nuclear white corpuscle, except in being a little smaller, more rigid, *i.e.*, less plastic and less adhesive ; qualities which may naturally be regarded as pertaining to a lower degree of development. Under the influence of re-agents both the primary lymph corpuscle and the uni-nuclear white corpuscle of the blood undergo a species of imbibition or osmose, by which a cell-wall is rendered very apparent.   This cell-wall is so delicate that its separation from the nucleus does not perceptibly diminish the diameter of the original corpuscle ; in other words the body is almost entirely made up of nucleus, and we must infer that the content matter lying between the cell-wall and the nucleus, which by the imbibition of water swells up so considerably must exist in a condition of great spissitude. If to cells of either kind in this swollen state saline aniline-blue be added, the corpuscles do not stain or undergo further change, even though they may remain in contact with the staining fluid for a long time ; but if by means of a suitable instrument the cover glass is firmly pressed upon and the corpuscles are simultaneously watched, they may be seen to burst with emission of the liquid which lies between the cell-wall and the nucleus, and he latter body becomes at once deeply stained.   The same kinds of changes are undergone by the multi-nuclear white corpuscles with the difference that from two to five nuclei are brought into view.   After the rupture has taken place, it frequently happens that portions of the unstained capsule and contained protoplasm remain attached to the swollen and stained nuclei.

Sometimes, when these corpuscles are compressed in the manner before described, instead of the external cell-wall giving way the wall or pellicle of the nucleus appears to yield, and its liquid contents are seen to pass out and to mingle with the general cell contents, and in this case (if staining fluid is present) the entire area of the corpuscle stains blue. Occasionally the content matter of the nucleus may be seen to bud out in processes into the general cell contents, and to remain some time without becoming diffused.   These facts seem to show that these corpuscles have not only a cell-wall but that their nuclei have at least something equivalent, if not to a wall, at least to a pellicular membrane.   Sometimes under conditions in which the cell contents seem less fluid than usual, and the corpuscles have been withdrawn from the Liquor Sanguinis by the " Method of Isolation, " this pellicle may be discerned.

Occasionally the pellicle or cell-wall of the white corpuscle may be seen in the absence of any special treatment.

Photograph 133, Plate XVII., shows three white corpuscles adhering to each other, and on their surface may be seen a crumpled pellicle which in the case of one corpuscle has become withdrawn for over a third of its surface, exposing the homogeneous content matter of the corpuscle beneath.

In Photograph 134 there are four white corpuscles, the envelopes of which appear to be unruptured, and the nuclei of which are obscurely seen through them, and not clearly as in Photograph 131, in which the pellicular covering appears to be absent. The tendency of late has been to regard the white corpuscles as consisting of nuclei and protoplasm only, without a limiting membrane. The appearances depicted in the photographs seem to be incompatible with such a view.

On the other hand, we have already given sufficient evidence of the possession of a cell-wall by the primary lymph corpuscles, and these corpuscles can readily be made to simulate uni-nuclear white blood corpuscles by keeping them for a time in some of the animal fluids, for example, urine or aqueous humor, and subsequently transferring them to fresh Liquor Sanguinis.

Photograph 126 shows primary corpuscles from the spleen of the pig which have been so treated. It will be noticed that they are considerably enlarged, and that large nuclei filling up their interior are brought into view. White blood corpuscles on the other hand can be made to assume more nearly the appearance of lymph corpuscles by transference to a quarter per cent. solution of common salt, as seen in Photograph 127, in which the dark outline common to some primary lymph corpuscles is restored. Photograph 125 shows the normal appearance in the same blood.

These observations, taken as a whole, seem to show that we have to deal in these bodies with true cells undergoing their ordinary mode of development, and they appear also to afford strong evidence that the uni-nuclear corpuscle of the blood is a *primary lymph corpuscle,* which, having failed to undergo the customary divarication necessary to its direct development into a colourless disc, continues its career as an ordinary cell, becoming more liquid and plastic, and acquiring the peculiar physical conditions which enable it to exhibit amœboid phenomena. At first a uni-nuclear corpuscle it increases in size, and as its development proceeds gives rise by the splitting up of its nucleus to the multi-nuclear varieties of the white blood cell.

It is not easy to trace experimentally the destiny of these nuclei; but analogy and reason conspire to indicate that these multinuclear blood cells ultimately disintegrate, and that their liberated and developed nuclei undergo conversion in the blood into *colourless discs,* which, gradually assuming colour, become red corpuscles. If this view prove to be correct, we shall have to recognize in the mammal two processes or modes by which colourless discs, which eventuate in red discs, are produced. (1) A major, rapid and direct one ; (2) A minor, slow and in-direct one, the former being superimposed upon the latter as a higher specialisation upon a lower type of development, in perfect harmony with the law of evolution, which regards the mammal as succeeding in the order of time and development the lower vertebrata. Diagram 1, Plate XX., and description will fully explain these two modes of development of mammalian blood.

SECTION VII.

On the Identification of the Advanced Lymph Disc with the Colourless Disc of the Blood.

WE have shown that the kind of development which the large majority of the lymph corpuscles undergo is not one which leads to increase of size or complexity of structure, as generally supposed (*e.g.*, the growth of a small uninuclear cell into a larger multinuclear one), but that it is, on the contrary, in the direction of simplicity and homogeneity—a simple nucleus derived from a uninuclear cell being converted into a smooth extremely delicate colourless disc. The unchanged *primary* lymph corpuscles when obtained from the blood-glands are a trifle smaller in diameter, but a little thicker than the red blood discs. They vary somewhat in size, but when not in an abnormal state not more than the red blood discs do among themselves. In brief, they are discs, which, like the red discs, have a slightly varying diameter, but at the same time a general equableness of size. These primary discs are comparatively rigid, probably owing to their furrowed corrugated state, for corrugation is found to effect a similar change in the red corpuscles. That the state of the lymph corpuscles is one of corrugation, and not of granulation, is made obvious by introducing them between glasses, which are close enough to slightly flatten them; they then increase a trifle in diameter, and become smooth, as in Photograph 102, Plate XV, which may be contrasted with Photograph 101, Plate XV. When in the smooth state their assimilation to the red corpuscles, save in colour and elasticity, is most striking. Many even of these *primary lymph discs*, as we have before pointed out, have a marked tendency to exhibit themselves in the biconcave form, and there is no doubt that this tendency becomes more and more established as the development of the corpuscles proceeds ; but the means at our disposal for showing this diminishes, because the corpuscles, by the loss of their envelopes, become more delicate, less visible, more fragile and fugitive ; indeed, altogether more incapable of resisting the changes which ensue in the environment when the lymph is shed. We cannot hope, therefore, to see these advanced nuclear discs in their most perfect form until, as with the colourless discs of the blood, we can devise some means by which they can be absolutely preserved. Osmic acid preserves exceedingly well the hæmoglobin discs of the blood, but it has less influence over the colourless discs, and less still over the advanced lymph corpuscles. The use of cane sugar has given better results than any other substance yet tried, and has enabled me to preserve both the primary and advanced lymph corpuscles in something like their true size and form. (*Vide* Photograph 111, Plate XV.)

I have before referred to the property which the advanced lymph corpuscles possess of staining and swelling up into discs of large diameter, under the influence of saline aniline blue, the most advanced corpuscles become the largest discs, and between them and the corpuscles which do not enlarge or stain, every intervening diameter may be found.

When we come over to the side of the blood, and investigate with the same staining fluid, we find corpuscles which swell up and stain precisely as the advanced lymph corpuscles, and others which stain only, and do not swell ; in other words, the power of this fluid to enlarge the corpuscles is lost before its power to stain them, and thus just as we are able on the lymphatic side to show a graduated series which culminates in the most advanced lymph corpuscle, so in the blood we can display the stages between the advanced lymph corpuscles and the hæmoglobin discs, for at a certain point when a little hæmoglobin has been attained by the corpuscles they cease to stain with the aniline blue.

The aniline blue, therefore, affords us a means of comparing the most advanced corpuscles of the blood-glands with the least advanced corpuscles of the blood.

In the healthy state of the organism it is probable that few corpuscles would leave the blood-glands save the most advanced ones, and these might be expected to undergo continuous development in the liquor lymphæ till they reached the termination of the thoracic duct, so that at this point we should expect to find the most advanced lymph discs, and therefore the ones which would most assimilate in properties and nature to the colourless discs of the blood, as found in the subclavian veins or the right auricle, at the time, and only at the time, that the thoracic duct is in the act of delivering up its contents to the blood. All the colourless discs found in the right auricle even at this period would not of course be advanced lymph corpuscles, inasmuch as colourless discs would be returned by the venæ cavæ. We are, therefore, more concerned in this connection with the state of the advanced lymph corpuscle at the thoracic duct, and with the products which flow spontaneously from the blood-glands, than with any elements we may be able to obtain by the incision of such organs, for the simple reason that in the latter case the products will be mixed and less advanced. Seeing the barely visible state of the advanced lymph corpuscles in the blood-glands, there can be no doubt that the great bulk of these bodies are in a condition as they leave the thoracic duct to become at once the invisible colourless discs of the blood ; but even this point need not remain a matter of speculation, for the method which I have devised for packing the red corpuscles together so as to display the existence of the colourless discs of the blood is also a method by which liquor sanguinis can be filtered off perfectly free from corpuscles, and into such filtered liquor sanguinis of venous blood it is quite competent for us to introduce advanced lymph corpuscles, and to compare them when placed

under these conditions with those which are found naturally in
venous blood taken from the right auricle during a period of
blood-gland activity.   This comparison has been made with the
result of showing that *the advanced lymph discs and the colourless blood discs are one and the same body.* We see, therefore,
how near to the mark the late Dr. Hughes Bennett was, when he
affirmed—" In chyle taken from the thoracic duct, there are
also biconcave flattened discs exactly resembling the coloured
blood discs in size and form, but destitute of colour." Now
this being so, where are these bodies a few minutes subsequently
to having passed into the right auricle of the heart ? Dr. Bennett recognised these corpuscles in the thoracic duct, but he
lost them in the blood, and filled up the hiatus by inferring that
they became coloured in the lungs, in their transit from the
right to the left heart. But even this should not have prevented
them being found in the right auricle, where by my methods
they can be detected in abundance, and also in the left heart,
and the circulation generally, which indubitably shows that
they are not lost by any *sudden act of colouration*, but are
simply rendered for a time invisible, as I have fully pointed
out, by falling into a liquid (liquor sanguinis) having the same
colour and refractive index as themselves.

    If it could be conclusively shown that the advanced lymph
corpuscles as found in the blood-glands, are colourless biconcave
discs of the same size as the red corpuscles, or that they can
become such before they leave the lymphatics (thoracic duct,
&c.), this fact alone would afford strong presumptive evidence
that they were the originators of the blood discs ; but when, in
addition to this, the discovery is made that colourless discs exist
in the blood in appropriate and corresponding numbers, and also
discs in every intermediate stage of colour between these and the
full red disc, it would seem that the claim put forth that the
red discs are but gland discs* in a further stage of development

---

    * In this term is included the products of the spleen, bone-marrow,
and the other blood-producing organs.

is impregnable, and the controversy as to the origin of these discs at an end. The remaining point relates to the acquisition of colour, and on this head we are only able to say that the change from a colourless to a coloured disc takes place in a gradual manner, and that it occurs while the corpuscles are circulating in the blood. We know, in addition, that a colouring principle exists in the blood which is not hæmoglobin, but which belongs to this series.* We may, therefore, reasonably ask the question, is it a function of the colourless disc to absorb and raise this colouring matter into hæmoglobin?

I have ascertained that the blood of the full-grown fœtus is very rich in ordinary white blood corpuscles and in colourless discs, but the latter are not nearly so fully developed and perfected as in the adult, but exhibit more the characters of discs taken directly from the blood-glands; indeed stronger proof could not be afforded that these colourless discs of the blood are the actual products of the blood-glands than is found in the fœtus. If we take corpuscles from an incision in a gland and add them to blood, we of course, introduce them into the blood in a much less perfect state than when they mingle with this fluid in the ordinary way, yet corpuscles so introduced compare singularly well with the colourless discs found by the ‘method of packing’ in fœtal blood. The difference between the appearance of the colourless discs of the fœtal and the adult blood is probably due to the excessive activity of the glands, in the former case, an activity akin to that present in leukhæmia in adult glands, the corpuscles are in fact the products of growing, and not of perfected glands, simply in the exercise of their functions. The same is true in leukhæmia, the growth, or hypertrophy of organs appears in these cases to be inconsistent with the perfection of the product.

Even the casual observer of the products of the blood-glands cannot fail to be struck with the protean forms and appearances, and the great varieties of size under which these bodies present

---

* *Vide* note, page 54.

themselves, and if the research be extended to the young
colourless discs of the blood, the same will be found true of
these also. In both cases this is due to their extreme adhesiveness
and plasticity, and to their great dependence upon the normal
state of the liquids in which they are submerged. The various
post-mortem changes of which lymph corpuscles are susceptible,
have been described on pages 15 and 105, and the changes which
the colourless discs undergo on pages 56 and 57. It will be seen
that the assimilation of these bodies in physical and chemical
properties is very remarkable. These have been discussed
under the headings of Size and Form—Colour—Adhesiveness
—Liquidity—Specific Gravity—Granulation—Enlargement—
Aggregation and Fusion—Spreading, Lamination and Fibrillation. It is obvious that with all these tendencies to
spontaneous change it becomes absolutely necessary to adopt
measures for the preservation of these bodies, and for the maintenance of their true form and size ; and unless this is done no
correct idea can be formed of their nature, or of the relation
which they bear to each other. When these precautions have
been taken we are able to demonstrate that no important
difference obtains between the advanced lymph corpuscle, as it
exists in the blood-glands, and the colourless discs of the
blood. In the first place they are both disc-shaped nuclei, of the
same diameter as the red corpuscle, and the former is frequently,
especially in young animals, irregularly and sometimes perfectly
biconcave, the latter, when well preserved, is usually smoothly biconcave, at other times more irregularly so, like the lymph disc.
In addition, these bodies undergo the same post-mortem
changes and stain with the same stains, the only difference
being that the colourless discs of the blood require a
stronger stain to display them, and do not become extended or
swollen so much as the lymph discs ; in other words as these discs
undergo their normal development, and depart further and
further from the lymph stage, they also undergo gradual
chemical alterations which are indicated by the refusal of the
corpuscle to become stained with a strength of aniline blue

competent to stain the lymph discs. They lose too their tendency to enlarge under the influence of this reagent, which is a marked property of the advanced lymph disc. There are, however, always to be found in the blood a certain number of discs which correspond in every respect with the advanced lymph disc. The extreme tendency of this chemical change is seen when the corpuscles arrive at that stage of their development in which they neither stain nor swell, and at this period they have acquired a moderate amount of hæmoglobin.

The origin of the red blood disc can be successfully traced back to the *primary cell* of the blood-glands, the disc-shaped nucleus of which is set free, becomes smooth and biconcave, enters the blood as its colourless disc, and gradually acquiring colour becomes its red corpuscle. If, on the other hand, the primary cell fails to undergo *enucleation* it passes on into the blood, and becomes there the uninuclear white corpuscle from which by proliferation of the nucleus, the multinuclear varieties of this body spring.*

---

* *Vide* Section on Leukhæmia.

SECTION VIII.

On the Rôle of the Red Bone-marrow in the formation of Blood.

It is the soft, reddish substance found chiefly in young subjects and growing animals, and which in the adult is confined mainly to the ribs and the vertebræ (not the fatty marrow), which is held to be the blood-producing organ. This substance is loaded with corpuscles, some of which appear to be identical with the white corpuscles of the blood, or with *colourless* embryonic cells, such as those from which the nucleated red cells of the embryo are produced, while others are like the primary gland and splenic corpuscles, and might readily be mistaken for naked nuclei. In addition to these bodies, there are to be seen others in the bone-marrow, which are structurally similar to the white corpuscles of the blood, but which possess a distinct yellow or red tinge. These are supposed to be intermediate products between white and red nucleated corpuscles. Such appearances as are here described are best seen in embryo bone-marrow, and in the adult state in that of some of the lower mammals, *e.g.*, the guinea pig.

It has been claimed by Neumann that in the bone-marrow of man and the hare one constantly finds cells resembling the ordinary white corpuscles of the blood, and nucleated cells which are distinct from the preceding cells, by the fact that they contain a yellow colour and have a more homogeneous consistence. These he considers to perfectly resemble those coloured embryonic blood cells which contain a nucleus. The nuclei of some of these cells, he affirms, become divided into several parts, and lose their clearness of outline ; and in some such cells he regards the nucleus as having entirely disappeared. These latter he holds to be the transition forms between the red nucleated cells and the red biconcave discs. These transformations he believes to take place within the *capillaries* of the bone-marrow.* According to this theory the red biconcave discs are formed in the bone-marrow from the red exteriors of nucleated corpuscles, the nuclei of which have been absorbed or have undergone atrophy.

Fourteen years have elapsed since this view was first promulgated, and during this interval, it has come to be freely admitted that the red bone-marrow is a source of colourless nucleated cells, and that it plays an important part in leukhæmia. yet it has never gained acceptance as a satisfactory explanation of the origin of the red biconcave discs.†

Professor Rindfleisch having been unable to satisfy himself of the existence of these transition forms from the colourless corpuscles, either in the bone-marrow or the spleen of the adult mammal, but nevertheless being inclined to admit the presence in the adult bone-marrow of a few red nucleated cells similar to those which are found more abundantly in the early embryo and a few of the lower mammals, proposes for these bodies the designation *Hæmatoblasts*. He considers that they multiply by sub-division as in the blood of the early embryo. In the very young embryo of the guinea-pig he has observed the nuclei of their red cells to become separated from the coloured protoplasm. This latter substance he believes to contract and become modelled into the red disc.‡ Rindfleisch freely admits that he has rarely seen this take place in the marrow of

---

* Many reasons have been given by Rindfleisch in favour of the view that no true capillaries exist in the bone-marrow, but the union between the arteries and veins is maintained by mere wall-less passages or channels. Under this view it is of course easy to understand how the bone-marrow products enter the blood.

† This is entirely in keeping with the fact that it has all along been easy to account more or less perfectly for the white corpuscles, but very difficult to form any notion as to the origin of the red disc.

‡ *Vide* Section I., Part IV.

the adult mammal, but attributes this to the paucity of such corpuscles, and to the area over which they are spread. This argument would not of course apply to the adult guinea-pig, in whose marrow the red nucleated cells are numerous.

As the views of Rindfleisch constitute the latest attempt to clear up the mode of origin of the red biconcave disc of the mammal, I shall refer to them at some length ; but before doing so I think it will be better to place before the reader the results of my own researches into the nature of the red bone-marrow. I propose first to inquire into the part played by the bone-marrow in the process of blood formation. To the extent that the morphological elements of the blood differ from each other, the question we have propounded becomes complex. Red bone-marrow may be derived from a variety of sources :—

1st.—From the mammal embryo during the early months of intra-uterine life.

2nd.—From the mammal fœtus at the later period of intra-uterine life, and at full term.

3rd.—From the young and the adult mammal.

4th.—From the oviparous embryo, during the state of incubation.

5th.—From the young and adult ovipara.

These bone-marrows require to be examined and studied in the first instance quite independently of each other, and also to be contrasted with the formed elements of the blood which exist at the particular period ; *e.g.*, during the very early period of intrauterine life in the mammalia the blood contains colourless and coloured nucleated cells only. At this period the formation of blood is purely intravascular, and arises from the subdivision of the colourless embryonic cells themselves. Later on the blood receives elements from without, but its corpuscles are still nucleated ones—white, red, and inter-mediately coloured ones (probably derived in the colourless stage from the spleen and lymphatics). A little later non-nucleated or red biconcave discs are added to these, so that the elements in the blood now consist of colourless nucleated corpuscles, red nucleated corpuscles, colourless biconcave discs,* and red biconcave discs of all shades. We will consider in the first place the relation which the bone-marrow of the mammal embryo bears to its blood, and subsequently that which obtains between the adult mammal bone-marrow and its blood.

On the Relation which the Bone-Marrow of the Mammal Embryo bears to its Blood.

If we examine the blood of the embryo during the early periods of intrauterine life, we find one of its constant con-stituents to be colourless, clear-bordered, nucleated cells. These cells are shown in great perfection in Photograph 149, Plate XIX. In some cases the clear hyaline border which surrounds the nucleus can be seen to have acquired a little colour; indeed, if we seek carefully we shall find that there are always present in this blood nucleated corpuscles of every shade of colour between the colourless one before referred to and the fully coloured nucleated corpuscle (*vide* Photograph 150 and 151, Plate XIX). The prior states of the clear-bordered, colourless, cell itself can also be traced in every stage of development in the blood from a nucleus-like looking corpuscle having an average diameter of about 6-20,000ths of an inch, but which in reality is a true cell, the capsule or wall of which is at first in close contact with its nucleus (*vide* Photograph 152, Plate XIX). This appears to be the analogue of the primary lymph corpuscle found in the lymph glands and in the spleen of the adult mammal, and its development into a colourless clear-bordered, nucleated cell, and from this into a coloured nucleated cell, can be traced with great facility in the blood of

---

* The invisible colourless disc or advanced lymph corpuscle.

the human embryo. In Photographs 150 and 151, Plate XIX., these colourless, clear-bordered, nucleated corpuscles may be seen lying among ordinary red discs, and the mode in which they gradually become enlarged and obtain colour may be traced. These observations appear to show that the clear, colourless, nucleated cells found in the blood of the embryo acquire their colour gradually in the blood; and this view is supported by the fact that these same corpuscles are found plentifully in the pulp of the embryo spleen, and have not in this situation acquired any colour. This may be seen in Photograph 153, Plate XIX., which is a specimen from the spleen of the same embryo. We see, then, that the red nucleated cell of the embryo commences its existence as the clear, colourless, nucleated cell, and that between the latter and the former every grade of intermediately coloured cells can be found ; but in addition to these cells, which constitute an unbroken series, there are present also in this blood, numerous ordinary red discs, and we find that these, too, have various shades of colour, from a colourless disc (invisible corpuscle) to the full red disc— precisely similar to what exists in all mammal blood, as depicted in Photograph 61, Plate XII. If from the embryo blood we turn to its bone-marrow, we discover in it many of the elements which we have already found to be constituents of the blood. When we reflect that the blood flows through mere wall-less channels in the red marrow, we may see how impossible it will be to take bone-marrow for examination without sometimes getting blood elements mixed with it, and it is clear that unless important differences can be shown between such elements as found in the bone-marrow and similar ones found in the blood, we should scarcely be justified in regarding them as belonging to the former source.[*]

---

[*] In order that erroneous ideas may be avoided as to the quantity and shape of the red marrow corpuscles, Rindfleisch advises that the vessels of the animal should be well syringed out with a three-quarter per cent. solution of sodium chloride, and then that the bone-marrow specimen should be diffused into a similar solution for examination. This plan has some advantages; but these are counterbalanced by the fact

The reader will do well to compare Photographs 150 and 151, Plate XIX., of blood obtained from the aorta of the embryo, and Photograph 154, Plate XIX., of bone-marrow from the rib of the same embryo. It is obvious that, taken by itself, there would not be sufficient difference between the red nucleated cells in the two cases to warrant us in affirming that those seen in the marrow had been produced *in situ* any more than there is sufficient difference between the red disc seen here and in the blood to warrant a similar conclusion. On the other hand, if there are any bodies found in the bone-marrow which are not found at all in the blood, or found plentifully in the former site and sparsely in the latter, we are entitled to regard such as forming its proper constituents. It appears obvious, however, that before we commence to discuss the function of the red marrow we must be in a position to deduct from it all such bodies as may reasonably be considered to be derived from the blood. In the blood of the embryo we have to account mainly for two elements—the red nucleated corpuscle, and the red disc. To do this we require to know from whence the blood derives its colourless, clear-bordered, nucleated corpuscle, and its colourless disc (invisible corpuscle). The antecedents of both of these bodies we can find readily enough in the spleen and lymphatic glands of the embryo, as the primary and advanced lymph corpuscles, and we can trace the development of both of them in the blood, right through their course, till they eventuate in red discs, and there seems also to be considerable evidence that some of the red discs found in the blood have resulted from changes in *its* red nucleated corpuscles, for similar proof of nuclear absorption can be found there as in the bone-marrow substance. (*Vide* Photographs 150 and 151, Plate XIX., and descriptions.)

---

that the bone-marrow cells swell up greatly under the influence of the saline. On this account I have always preferred to examine these bodies in their own plasma. No doubt syringing out the vessels enables us, as Rindfleisch says, to be more certain that what we have before us has really formed part of the true bone-marrow substance.

We must now inquire what elements the embryo bone-marrow contributes to the blood, and in what condition these enter it.

The embryo bone-marrow has this peculiarity, that certain of its constituents are possessed by it in common with the embryo blood. Thus it has the primary lymph, and the colourless, clear-bordered, nucleated corpuscle, which originates from this by osmosis.* This corpuscle exists in all stages of colour to full red. The colourless or invisible discs and red discs exist also in all stages of colouration, with certain diversities of appearance. All the elements we have just mentioned can be found in the embryo blood, and therefore the temptation to view them as accidental contaminations from this source is very great, for everyone must feel how difficult it is to obtain specimens of bone-marrow for examination entirely free from blood.

If we were alone concerned with embryo forms we might readily accept this view, and by the withdrawal from consideration of the elements which are also present in the blood, place the bone-marrow on a level with the spleen and glands as yielding only *colourless products ;* but we have before us the fact that in the guinea-pig these red nucleated constituents are present in the bone-marrow throughout *adult life*, and become fully coloured *in situ*, and the additional fact that no *nucleated coloured cells* can be found in this blood. Now, if none of the fully developed, red nucleated elements of the adult guinea-pig pass on into the blood, it is unreasonable to suppose that in the normal order of things any of the corpuscles which constitute the antecedent stages of these bodies do so. As we are in a position to indicate with great accuracy in the guinea-pig the colourless bone-marrow corpuscles from which the coloured nucleated ones arise, it is clear that we must also regard these as stationary elements of the marrow. If the marrow of the adult guinea-pig yields any products to the blood by virtue of the presence in it of red nucleated cells, these must be fully formed red discs,

---

* *Vide* page 117.

and we may expect, therefore, to find a few in the bone-marrow, giving more or less evidence of being recent formations. The whole series of elements or corpuscles concerned in the production of the red nucleated corpuscles of the adult guinea-pig must necessarily be stationary products of the bone-marrow, and can have no other purpose besides the production of red discs through the intermediary agency of red nucleated cells.* This being the case it is probably true also of the embryonic condition; indeed, this process in the adult guinea-pig, and some few other mammals, appears simply to be a persistence throughout life of conditions which are essentially embryonic. Further evidence that the embryo bone-marrow is to be regarded as a region in which blood corpuscles are produced, and coloured *in situ*, is found in the fact that side by side with the process

---

* Rindfleisch appears to have been unable to trace the connection which subsists between the colourless cells of the bone-marrow and the red nucleated ones, for he says, "I have only negative information to give on the subject of possible prior states of the red nucleated cell. In the course of my numerous attempts to find out the right mode of development I have been obliged continually to turn away from the probability that there is a granulous prior state of the red nucleated cell. There are granulous cells which older creatures possess in some quantity in the bone-marrow, but these are mostly regressive elements or pigment cells in the marrow, but not young blood corpuscles. That colourless blood corpuscles gradually get a reddish yellow fringe, and thus necessarily change into red nucleated corpuscles, is an hypothesis which I share with my fellow-professors, and which I uphold at present, but which I cannot sufficiently prove from examinations made on guinea-pigs, or on other animals who have non-nucleated red blood corpuscles. I shall later on state that among the colourless corpuscles of birds' spleen the required forms of transformation are found. I have not made identical observations from the marrow and spleen cells of the guinea-pig, hare, pig, or man. It appears to me likely that there exist certain transformations from the marrow cells to the red nucleated cells, but as the necessary proofs are disproportionately few, I had to leave this hypothesis as possible, and to content myself with the positive facts that the larger red nucleated cells increase by division, and that the younger cells are about ⅔ths the size of the mother cells." The detection of the transition forms depends almost entirely upon the capacity of the individual to estimate the presence of delicate shades of colour, and under some circumstances this becomes impossible. It is then that the exquisitely delicate test afforded by photo-chemistry comes to our aid.

which forms red discs from red nucleated cells, an entirely different mode of red disc formation is simultaneously going on which corresponds entirely to the major mode of blood development in adult mammals, and to which I have in this work devoted so much consideration, viz., *the formation of red discs by the direct colouration of nuclear bodies.* This process, which may be seen in Photograph 155, Plate XIX., in full operation, was going on in another rib of the same embryo from which Photographs 150 and 151, Plate XIX., were obtained. This differs only from the general method in which mammalian red discs are formed in the fact that instead of occurring in the circulating blood, it is taking place in a special locality through which blood freely permeates, and the lymph nuclei are seen to be taking up colour before they have attained that exquisite delicacy and smoothness which they possess as the invisible corpuscles of the blood, *i.e.*, they still possess the furrowed irregular surface peculiar to lymph nuclei.*

If now we turn our attention to the colourless cells of the embryo bone-marrow, we find them to be present in large numbers, and to exhibit very diverse appearances. Thus we have—

1st.—Smooth dark cells of variable size, ranging in fact from 6 to 8-20,000ths of an inch in diameter.

2nd.—Cells which appear to be the same as the previous ones, but which look larger, owing to their envelopes and dark protoplasm being now in a coarsely granular state. From many of these cells this coarse, outer covering is obviously undergoing removal, and exposing a lighter coloured, more finely granulous interior body. (*Vide* Photograph 156, Plate XIX.)

3rd.—A free nucleolated nucleus, which previously lay within the coarsely granular cell, but which is set at liberty by the dissipation of the exterior dark granules. (*Vide* Photograph

---

* This mode of blood formation in the embryo has hitherto escaped observation, and is strong corroborative evidence of the general views maintained in this work.

157, Plate XIX., in which the smaller corpuscles are the ones referred to).

4th.—In the advanced mammal embryo, when ordinary red discs are being formed, we find in addition, by the aid of saline aniline blue, which swells it up into a large disc, as in the case of the advanced lymph corpuscle, a free nucleolus, which has been set at liberty by the disintegration of the capsule of the body number three. These bodies when stained swell up greatly, and show themselves as in Photograph 112, Plate XV.

In the section " On the morphological elements of the blood glands," I have shown that the earlier or primary cells have a pigmentary character, and are sometimes, when photographed, as dark as the red corpuscles. Examples of this may be seen on reference to Photographs 103, 108, 110, 116, etc., Plates XV. and XVI. The same thing is true of the early cells in the bone-marrow, and, as in the case of the lymph corpuscle, this dark protoplasm undergoes disintegration, and probably solution. The original bone-marrow corpuscle appears to be a perfectly typical cell, i.e., a structure possessing an external wall, a considerable amount of pigmented protoplasm, a nucleus and nucleolus, and the changes which take place in it, for the production of the blood corpuscles, nucleated or non-nucleated, are of a regressive character, the disintegration of its primary capsule giving rise to the body from which the clear colourless nucleated corpuscle is produced, and of its secondary capsule the body from which the non-nucleated mammalian disc originates.

It would appear that in most cases pigmentary matter is brought with the corpuscles, probably as the basis material for the formation of hæmoglobin.* In the spleen and gland corpuscles of the adult mammal this is got by the disintegration of the capsule and protoplasm of the primary corpuscle, which sets free the nucleus necessary for the formation of the invisible disc, but in the case of the mammal embryo and of the ovipara,

---

* *Vide* page 107.

where a nucleated corpuscle is demanded, a more typical cell is required, *i.e.*, a cell possessing a nucleolus, and thus the outer layer of pigmented protoplasm can granulate off as usual into the plasma of the bone-marrow.

In Photographs 158 and 159, Plate XIX., of bone-marrow from the sternum of the *adult* guinea-pig, we are able to trace all the transition forms from the coarsely granular, pigmentary cell to the red nucleated cell, and we notice that although cells of all sizes are present, yet that there is a general tendency in corpuscles as they assume colour for the nucleus to become smaller and the cell more condensed, approaching more nearly to the size of the red discs; but although the nuclei have become small and comparatively smooth, they are still distinctly visible, and if the condensation or atrophy of the nuclei proceeded much further, it seems probable that they would in some instances become even smaller than the red discs. There appears also to be evidence that the altered and modified nucleus may disappear ultimately from view by itself becoming coloured. There is nothing extraordinary in this conception when we reflect that this nucleus is the very body which in the second mode of corpuscle production in the embryo, becomes coloured after it has been freed from the colourless cell of which it formed a part.*

---

* Rindfleisch considers the red disc to be formed from the red nucleated cell by the *emigration* of the nucleus of the latter. He has observed this separation to take place in very young embryos 5mm. in length, but has not seen it in older embryos, or to any extent in adult mammals. For my own part I regard it as a purely accidental circumstance not due to the emigration of the nucleus, but to the fact that the corpuscle becomes adhesive to the slide at the point where the nucleus lies, and the soft red protoplasm is then torn away from it by currents in the liquid. It is in this way that the temporary balloon or bell-shaped forms are produced. If these forms or approximations to them are retained after the act of separation, it is owing to the elasticity of the corpuscle being held in abeyance by the altered state of the serum. Frequently, when this occurs, the separated mass of coloured protoplasm assumes at once the globular form. My own observations of the manner in which red discs are formed from red nucleated cells in the embryo and in some of the lower adult mammals are more in favour of the absorption and modification of the nucleus *in situ*.

The cells in the adult guinea-pig which become coloured are very variable in size, and the nuclei in them are often quite incapable of being seen, and this both before and after they have attained colour. This appears to be due to the density and opacity of the protoplasm, and it seems probable that if it absorbed a certain amount of water it would become transparent, and the nucleus would be disclosed.

There are small coloured corpuscles in the marrow of the adult guinea-pig which look very much like coloured nuclei; but I think a close inspection of these bodies under favourable conditions will invariably show them to be nucleated cells, possessing a minimum amount of protoplasm. I allude to the small coloured corpuscles which adhere together in groups, as shown in Photograph 160, Plate XIX. As one is not able to trace in the bone-marrow of the adult guinea-pig the second mode of blood formation, such as is seen in the embryo, viz., the direct conversion *in situ* of colourless nuclear discs into red discs, it is therefore probable that this mode, if it obtains at all, goes on to a very limited extent only, having given place to the regular mode which obtains in the higher mammals, viz., to that of producing and transmitting colourless non-nucleated discs to the blood. For further remarks on this point see description to Photograph 160, Plate XIX.

On the Relation which the Bone-Marrow of the Adult Mammal bears to its Blood.

Previous writers on the bone-marrow have endeavoured to show that its function in the adult mammal is to produce a red nucleated cell * which undergoes transformation into a red disc.

---

* Rindfleisch has given the name *hæmatoblast* to this red nucleated cell, while Hayem has proposed the same name for the colourless, clear nucleated cell, which he regards as the antecedent of the red nucleated corpuscle in the ovipara.

This idea is essential to their conceptions of the mode by which this substance contributes to the development of the blood.

This view has been largely derived from the study of embryonic bone-marrow, and from the fact that red nucleated cells could always be found in the marrow of some adult mammals, *e.g.*, the guinea-pig. In the presence of this view, it becomes a most essential point to ascertain two things.

1st.—Whether red nucleated cells exist in the bone-marrow of all mammals.

2nd.—If they do, whether they exist in sufficient numbers to exert any important influence over the production of the red discs.

As the result of a most rigorous and protracted inquiry into this question as to whether all adult bone-marrows contain red nucleated cells as a constant and essential constituent, I have been able to arrive at the following conclusions :—

In certain mammals, such as the guinea-pig, rabbit, etc., red nucleated cells can be found in fair quantity, even at an advanced period of life.

In large mammals, a few weeks old, *e.g.*, the calf, a few aborted structures* of this kind may occasionally be seen. On the other hand, the most rigid scrutiny of the bone-marrow of the ox, adult pig, sheep, etc., has failed to discover, except at the rarest intervals, even aborted bodies of this kind. Now, it is obvious that, if all mammals are equally dependent upon these bodies for the formation of their red discs, they should in all cases bear a definite proportion to the number of red discs produced. If, for example, they bear the same share to the development of the red discs of all mammals as they do to those of the guinea-pig, then they should be present in the bone-marrow of such mammals to the same extent as in that of the guinea-pig, and, if this is not so, it shows that what is taking place in these lower mammals is altogether exceptional

---

* Structures altogether incapable of forming red discs either by absorption of the nucleus or separation of the imperfectly coloured and scant exterior.

in its character. In my observations on the bone-marrow of
the embryo I have stated that structures which are only
sparsely found in an organ must be viewed with considerable
suspicion, as probably being accidental contaminations, or as
being the products of functions or processes which have undergone incomplete suppression. The latter view, is, I believe, the
true explanation cf the presence, in moderate numbers, of red
nucleated cells in the bone-marrow of young and adult lower
mammals, and of their almost entire absence in that of the
higher.

The embryonic mode of red disc formation from uncleated
cells gives place more or less completely to the more rapid and
less roundabout processes which I have shown even in the
human embryo, to be superimposed upon the former, viz., the
formation of red discs by the colouration of free nuclei.

If after the examination of the bone-marrow of the
embryo mammal, or of such marrows as those of the
adult guinea-pig, we turn our attention to this substance
in the higher mammals, we are struck with the great
paucity of coloured blood elements, the marrow appearing
to consist almost entirely of similar colourless cells to those
which we have described in the bone-marrow of the embryo.
Thus we have no difficulty in finding the smooth, opaque, dark
cells which, by the granulation of their protoplasm, give rise to
the coarse, dark granule cells. Photographs 161, 162, and 163
(from the rib of the ox), and 164 (from the pig), and 165 (from
the rabbit), Plate XXI., give examples of these bodies in their
smooth and granular condition. This dark capsule can constantly be seen granulating off, and it then reveals a lighter
comparatively opaque body, in the interior of which, generally
hidden from view, there lies a nuclear structure (*vide* Photographs 161, 162, 163). In some of them which have become a
little more transparent this nucleus, or rather original nucleolus,
of the perfect cell can be readily seen (*vide* Photograph 166).
The exteriors of these cells appear in their turn to disintegrate,
and to set at liberty *naked nuclei*, which appear in some instances

to gain a little colour while still in the bone-marrow, but which in the main pass on into the blood to swell the number of its colourless discs (*vide* Photographs 167 and 168, etc.) As in the mammal embryo, so in the adult, the colourless corpuscles of the bone-marrow, various and diversified as they seem, appear to result entirely from changes in a typical cell, which is present in various stages of growth and development, and which, being exceedingly prone to *post-mortem* change, becomes much swollen and altered from its normal states.*

In the bone-marrow of the higher mammals it is quite obvious that the embryonic mode of producing the red disc has entirely given place to the splenic and lymphatic modes, and there is no longer any difference in the colourless products of these organs, except that in the bone-marrow, the cell from which the white corpuscle and the red disc is produced, is seen a stage further back in its existence, viz., in the condition of a smooth and coarsely granulous pigmentary cell.† The nucleated bodies which are set free from these dark masses of granules, correspond entirely to the cells I have in other cases designated the primary lymph corpuscles, and can, as in their case, be made by osmosis to show their external envelopes, while the nuclei at once stand revealed (*vide* Photograph 169, Plate XXI., from bone-marrow of calf). By the addition to the *fresh* marrow of any mammal of very weak saline aniline blue, the nuclei which are already set free are brought palpably into view. (*Vide* white-spreading corpuscle in Photograph 170, Plate XXI., and its description.)

There seems, therefore, to be in the mammal (embryo and adult) two methods by which non-nucleated red discs are produced :—

1st.—By the conversion in the bone-marrow of red nucleated cells directly into red discs. This method is in operation in the mammal embryo generally, and also in the adult guinea-pig, rabbit, etc.

---

* *Vide* Page 15, Section I.

† In the calf these pigmentary coarsely granular cells appear to be fewer in number.

2nd.—By the colouration of naked lymph nuclei. This is the super-imposed or advanced mode of production of red discs in the mammal embryo, and may be readily seen in operation, because the lymph nuclei are gaining colour *in situ*, and are less developed, *i.e.*, more like lymph corpuscles than they are after they have become the colourless discs of the blood. (*Vide* Photograph 155, Plate XIX.)

The only difference between the second mode of development in the embryo and that which goes on generally in the blood of the adult mammal, is that the lymph corpuscles become coloured in the former case in the bone-marrow, but not in the latter case till they become denizens of the blood, and by this time they have lost their corrugated appearance, have become smooth, and are no longer to be seen under the ordinary methods of examination. The few exceptional instances in mammals in which fully developed red nucleated cells are found in the bone-marrow, and probably give rise to red discs, are to be regarded as due to the survival or persistence of the embryonic method, which is gradually giving place to a higher mode, but retains its hold in the bone-marrow for a longer period than elsewhere. Even in these cases of the lower mammals where red discs are being formed in the bone-marrow alone, in small numbers, from red nucleated cells, the same bone-marrow furnishes to the blood advanced lymph discs, to become converted in the blood into red discs by gradual colouration. Some mammals are exceptional (so far as their bone-marrow is concerned) in retaining traces of the embryonic method of development. The embryo mammal developes its red discs also from the advanced lymph corpuscles, and its nucleated red corpuscles from the primary corpuscles of the spleen and lymphatic glands. Both these corpuscles get their colour in the blood.

In the higher mammals, shortly after birth, the development of red dics from red nucleated cells is almost entirely suppressed, and the bone-marrow, like the spleen and the lymphatics, yields to the blood free naked nuclei, which become transformed into its colourless discs, and gradually by the attainment of colour into red discs. In the higher mammals, therefore, the tendency is to complete uniformity of method to the extent that the embryonic conditions of the bone-marrow are departed from.

Rindfleisch admits that it is in the guinea-pig bone-marrow that he has had most success in finding coloured nucleated cells, and also that their spleen products will not help out this view of blood development. The guinea-pig has not a greater number of red discs than other mammals, and if in all mammals they are produced from red nucleated cells, then these latter cells should be formed in as great numbers in their bone-marrow, whereas in most cases they are almost entirely absent. The fact is, the embryonic mode of blood development in the bone-marrow from red nucleated cells does not persist in the higher mammals, save as an aborted and suppressed function.

Conclusions.

The preceding observations warrant, I think, the following general conclusions :—

1st.—That at an early period in the life of the mammal embryo, red nucleated cells are developed from the colourless corpuscles of the bone-marrow.

2nd.—At a more advanced period of embryo life the red nucleated cells appear to undergo in the bone-marrow conversion into red discs, and simultaneously a new method of development of red discs springs up, which consists in the assumption of hæmoglobin by naked nuclei which have been set free by the disintegration of some of the colourless cells of the marrow.

3rd.—After birth, in the majority of mammals, the formation of red discs by the modification of red nucleated cells ceases, or becomes an aborted process, and the formation of red discs from naked nuclei is so far modified that the free nuclei leave the bone-marrow and enter the blood in a colourless state to acquire colour in the circulation.

4th.—In some few mammals the embryonic mode of forming red discs by the modification of red nucleated cells continues to a limited extent, throughout life, alongside the numerically far more important one of transmitting naked nuclei to the blood to become its colourless discs.

5th.—As in the splenic and lymphatic forms, bone-marrow leukhæmia arises from an excessive formation of colourless nucleated cells, which are forced over into the blood before they have undergone those regressive changes which fit them to become the colourless discs of the blood.

On the Nature of Leukhæmia.

As everyone knows, the disease which bears this designation was discovered in our own time almost simultaneously by Bennett and Virchow. The attention of these observers was arrested by the presence in the blood, of certain anæmic-looking patients, of an unusual number of white blood corpuscles. The latter author subsequently showed that this condition of the blood was associated with enlargement of the spleen and of the lymphatic glands, and this observation has strengthened the opinion that these are the organs which normally supply the blood with its white corpuscles.

As blood highly charged with these colourless bodies becomes much paler in tint than usual even to the naked eye, the application of the term white blood is justified. Further observation of this disorder has shown that the red blood discs diminish in the ratio of the increase of the white corpuscles. The reason of this has received no explanation, and according to the view generally held that the red discs originate only from the white corpuscles it would seem to be inexplicable ; for the greater the number of progenitors the more numerous one would think should be their offspring.

It is generally affirmed that the condition of the spleen and of the lymphatic glands is one simply of enlargement or hypertrophy, and although these organs may attain to a very great increase in volume, even in the case of the lymphatic glands, nevertheless the condition is simply one of a true over-growth, *i.e.*, of a proper relative development of all the histological constituents. If this is so we can look for nothing but an increase in the elements which it is the normal function of the organ to produce, and if the assumption be correct that it is the function of these organs to produce bodies identical with the white corpuscles of the blood, then the excess of these bodies present is easily understood, but the difficulty in respect to the origin and development of the red discs is increased. In Section V., " On the morphological products of the blood glands," I have endeavoured to show that the true physiological function of these organs is not to produce spherical cells analogous to the white blood corpuscles, but, on the contrary, *nuclear discs*, which in the ordinary course of things take their place in the blood as its *colourless biconcave disc.*

I have further shown that the process by which this is effected is by no means roundabout or slow, but, on the contrary, short and direct; that, in fact, a more rapid mode of blood production could not be readily conceived ; and that to the extent that this method is departed from, the blood-making process in the mammal becomes a comparative failure; in other words, the process of blood-making is perfect in proportion to the extent to which the mode of development which I have designated the *major process* is in operation.

In our efforts to understand the nature of leukhæmia, we must keep steadily in view the fact that the bodies which it is the function of the blood-glands to produce are discoid ones of about the same diameter as the red blood-discs which are divisible into two kinds, representing stages in their development, a primary one or true cell, and an advanced one, which is the disc-shaped nucleus of the former. We must also recognise the fact that the blood-glands are not simple sites for cell production, but also regions in which a process of development occurs; that, in fact, the gland reticulum subserves the function of a delay station or incubating ground, in which the conversion of the primary into the advanced lymph corpuscles takes place. In other words, these trabecular networks are regions of arrest and entanglement, in which the

corpuscles are detained while being submitted to the processes
which determine the disintegration of their external envelopes
and the liberation and development of their contained nuclei. As
this transformation occupies some time, a certain length of stay
of the corpuscles in the retiform tissue of the glands is obviously
necessary. How the primary lymph corpuscles originate in the
blood-glands we are not at present in a position to state; but
from a careful consideration of the anatomical constitution of
the lymphatic glands we may obtain the assurance that their
production takes place in the central portions of the pulp
cords in immediate contact with the capillary vessels, and that
the corpuscles first produced are pressed outwards towards the
lymph sinuses and paths. From this, it follows that the alterations which are effected in the corpuscles during their residence
in the glands occur principally while they are passing from the
centre to the periphery of the pulp cord to be washed away in
the lymph stream. The time, therefore, allowed for development in the glands is determined by the length of time this
process takes, and there is, no doubt, a particular rate which
gives the most perfect results.

Now it might seem, that if we had slow and uniform
enlargement of the glands, as is the case in leukhæmia, with
regular and well regulated increase of all its constituent vessels,
cells, reticula, lymph paths, etc., then there would be no reason
why the primary cells should not remain the normal length of
time and undergo their customary conversion into advanced
lymph corpuscles or nuclear discs, and in such a case there
would be no leukhæmia, because it would simply be equivalent to
the substitution of a larger gland for a smaller one, the function
remaining unaltered; hence it would seem that mere hyperplasia
is not competent to the production of a leukhæmia. It is
necessary, however, to bear in mind that the process of cell
formation, as it occurs in these glands, is an intermittent one,
dependent upon those natural periodic hyperæmias which occur
at considerable intervals in connection with the digestive
process, and that between these periodic accessions of blood

there is a time of comparative quiescence, in which little or no fresh cell production occurs, but which may fairly be regarded as a period devoted to cell elaboration, and to the development of the corpuscles which are destined to be thrown off at the next period of turgescence. This is the normal process.

On the other hand, in leukhæmia the glands enlarge and grow under an excitement derived from a pathological stimulus too mild to excite in them a state of actual inflammation, but capable of maintaining a condition of constant hyperæmia, and of tissue activity sufficient to lead to a continuous excessive production of the primary lymph corpuscle ; in other words, the process of cell formation will be rendered continuous with periodic exarcerbations. Such pathological conditions would inevitably limit the length of stay of the lymph corpuscles in the glands, and in this way interfere with their due development, and cause them, instead of becoming converted into free nuclei to pursue a cellular development and to present themselves in the blood as uninuclear and multinuclear white corpuscles, as I have before explained in treating of the origin of the white blood corpuscle. *In this way we are able to understand how a constant hyperæmia, which leads to hypertrophy, also leads at the same time to numerical increase of the white corpuscle at the expense of the red.*

Abundant evidence has been given in the course of this research to show that the lymph corpuscle and the white blood corpuscle cannot be regarded as *identical*, and that the early form of the lymph corpuscle is a body *sui generis*, which may be converted, according to the conditions in which it is placed, on the one hand into a colourless biconcave blood disc, or on the other into a white blood corpuscle.

The nucleated white corpuscles, as seen either in the thoracic duct or in the blood in the healthy state, or in excess in the diseased condition leukhæmia, are to be regarded as primary lymph corpuscles, which are pursuing the ordinary course of cell development, because they have failed to remain long enough in the gland pulp to lose their cell-walls and become free nuclei as a preliminary step to their conversion into colourless biconcave blood discs. *The essential fact of leukhæmia is the failure on the part of the glands to properly set free and develope the nuclei of the primary lymph corpuscle.* For every white blood corpuscle formed, the production of a colourless disc, and therefore of a red corpuscle, is, to say the least, postponed, and a slow method of blood formation is substituted for a rapid one. We see, then, how simple a matter it becomes on this view to account for the increase of the white corpuscles of the blood and the coincident decrease of the red ones.

I repeat that the blood glands are not, as hitherto universally held, to be regarded as the breeding ground for the white corpuscles of the blood, but, on the contrary, for *the colourless discs which are destined to become the red corpuscles,* while the white corpuscles are in truth but a small minority of primary lymph cells which have escaped the changes necessary to fit them to follow this course. It will be seen that leukhæmia is a condition in which the latter process becomes excessive at the expense of the former : in brief, it consists in the persistent and continuous production of primary lymph cells at a rate which enforces their exit from the glands before they have been in them long enough to undergo their normal development into free nuclei and to become perfected into advanced lymph corpuscles, in which case they grow and run their course as white blood cells, or pass into the blood as imperfectly developed nuclei, instead of being transformed into colourless and subsequently into red discs. *Leukhæmia, in a word, is the encroachment of the minor upon the major process of blood making.* (*Vide* Diagram 1, Plate XX., and description.) As all the sources of blood production (lymphatics, spleen, and red bone-marrow) may become involved in the leukhæmic dyscrasia we may have one, two, or all of those forms represented in the blood, which have been specially designated lymphæmia, splenhæmia, and myelhæmia. The first is probably distinguish- able by the colourless corpuscles being a little smaller and having more simple and more strongly granular nuclei. The ordinary large, finely-granulous, white corpuscle may be taken as evidence of the second ; while the third variety is more liable to yield to the blood a *coarsely* granulous colourless corpuscle. All these bodies will be nucleated cells. On the other hand it is desirable to bear in mind that with such nucleated cells there may pass over into the blood many *free nuclei* in an imperfectly developed condition ; for the function of the blood-glands is not simply to set free the nuclei from the primary lymph cells, but also to develop and fit them to enter the blood as its colourless discs. This often fails to be accomplished in leukhæmia, and as a consequence many free disc-shaped nuclei which have failed to become smooth and biconcave present themselves in the blood. The precise condition present in any case may always be readily detected by means of weak saline aniline blue, which distinguishes the free nuclei from the cellular bodies, staining and swelling up the former in proportion to their degree of development, and, on the contrary, failing to stain and tending to cause contraction in the true cells (*vide* Section V.) It is desirable also to bear in mind the *possibility* of a pseudo-leukhæmia arising in the blood itself, independently of any derangement of the sources of corpuscular supply ; for if we reflect that the blood contains in the shape of colourless discs a great amount of colourless matter, which ordinarily is not seen, but which might be brought into view by any morbid condition of this fluid, capable of causing the colourless discs to aggregate into masses and to become granular, we can easily understand how an embolic and highly fatal form of disease might rapidly arise in the course of other morbid states.

SECTION X.

On the Nature of Anæmia.

The designation anæmia is perhaps more applicable to that condition which obtains after severe hæmorrhage, in which the volume of the blood lost is replaced by water abstracted from the tissues, a condition in which *all* the solid constituents of the blood are proportionately diminished. The term has, however, received a much more extended application, and is now used to denote conditions in which the red corpuscles only are deficient in number, and are replaced by water. In a given volume of such blood, say 1000 parts, there is usually a decrease of the solids and a corresponding increase of water. Most anæmias therefore involve hydræmias. The essential feature of anæmia, arising from disease, consists in a *deficiency of the average amount of hæmoglobin per corpuscle*, and generally is, though exceptions are said to exist, associated with a numerical deficiency of the red corpuscles.

It is to be observed, however, that this deficiency of hæmoglobin is usually out of all proportion to that of the red corpuscles,* and is, therefore, but partly explicable by their diminished numbers, at least under the ordinary concep-

---

* When blood is lost by hæmorrhage, the hæmoglobin average per corpuscle is not diminished, hence the loss of corpuscles and of hæmoglobin is equal and proportionate, *i.e.*, if half the corpuscles were lost, half the hæmoglobin would be lost also. In contrast to this, we find that in anæmia arising from disease a reduction of the corpuscles to one-half is frequently associated with a reduction of the hæmoglobin to one-fourth, involving a reduction in the average per corpuscle to one-half its normal amount. The cause of this singular fact has hitherto received no explanation.

tions as to the manner in which the hæmoglobin is distributed among them.

It is not intended in this essay to do more than indicate the novel interpretations of previously ascertained facts, which the new conceptions of the nature of the blood elicited by this research appear to render possible and legitimate. The points which appear to have special bearing on the subject of this paper, are as follows :—

1.—The corpuscles destined to become red corpuscles enter the blood as *colourless discs*, and become coloured while circulating in it, by a gradual, regular, and continuous assumption or production of hæmoglobin.

2.—The amount of fibrin in the blood may (in the absence of leukhæmia) be always taken as indicating the number of the colourless and slightly coloured discs which are present, and therefore the existing activity of the blood-glands.

These now well-ascertained facts form a secure basis for further considerations affecting the main issues we have to consider, viz., the simultaneous, but disproportionate decrease of hæmoglobin and red discs.

It is obvious that the numerical maintenance of the *red corpuscles* in their normal state demands—

1.—That the renewal and destruction of these corpuscles shall, taking the average, occur at one and the same rate.

2.—That the normal rate at which hæmoglobin is formed in the corpuscles shall be maintained.

3.—That, taking the average, each corpuscle shall possess the same term of existence. In a word, the maintenance of the normal corpuscular balance in the blood involves equal rates of supply and destruction, equal rates of colouration, and equal periods of existence.

Bearing these facts in mind, we will turn our attention to chlorosis and anæmia, and ask ourselves in the first place how the decrease in the red corpuscles is to be explained.

In this connection several conceptions are possible :—

1. We may suppose the functional activity of the blood-

glands to be *depressed*. This would give rise to a numerical deficiency of the *advanced lymph corpuscles*, and, as a consequence, fewer colourless discs would pass into the blood. Such an action would reduce the fibrin below the normal. This is ascertainable by analysis. Assuming the rate of colouration to be normal, the corpuscles, although fewer in number, would obtain their maximum colour, and, therefore, would yield the normal average of hæmoglobin per corpuscle, hence it is clear that these are not the conditions which obtain in ordinary anæmia. There would be in this case a deficiency of red corpuscles, and consequently of the total amount of hæmoglobin ; but there would be no deficiency of the average amount of hæmoglobin per corpuscle, as in anæmia.

2.—We may consider that in consequence of digestive disturbance, and impaired nutrition, the colourless discs, although entering the blood in normal numbers, may possess a lowered vitality, and, as a consequence, a diminished capacity for existence, and, therefore, be unable to live out the period necessary for the attainment of their maximum degree of colouration.

In such a case, although they might form hæmoglobin at the normal rate, they would cease to exist before they had acquired their full colour.. This view is therefore competent to explain both the loss of corpuscles and of hæmoglobin, and, as we shall presently see, the reduction of the average of the latter per corpuscle.

3.—We may hold that the supply of corpuscles is normal, that they are perfect in character, and form hæmoglobin at the normal rate, but that they are *destroyed* more rapidly than usual. This too, would explain both the loss of corpuscles and the average loss of hæmoglobin per corpuscle. Although the idea is in this case different to that in the former, the results would be *practically* the same.

4.—We may suppose that the supply is numerically normal, but the vitality lowered, and the rate of colouration diminished also. In this case the early death of the corpuscles will diminish their number, at a certain ratio, and the hæmoglobin average will be lowered, both by this and by the diminished rate of colouration. These two factors will operate simultaneously in the reduction of the hæmoglobin.

5.—We may consider that the colourless discs which enter the blood are normal, both as to number and state, but that the material upon which they react to form hæmoglobin is deficient in quantity or quality, and that as a consequence the colouration is imperfect. In such a condition there would be a greater number of fugitive discs, and therefore a corresponding increase of fibrin. In addition, we ought not to be able to count the usual number of corpuscles, for a greater number would be in a nearly invisible state, and those which had but a small amount of colour would readily be deprived of it by the diluting fluid, and a greater number of colourless ones would continue in the uncoloured and invisible state. To this order belong the cases in which the corpuscles remain nearly normal as to numbers, while the hæmoglobin becomes less than normal.

6.—It is conceivable that owing to impaired nutrition the colourless discs may have a qualitative defect, which causes them to form hæmoglobin at a slower rate, although the material for such transformation be present in normal amount. The result, as far as the corpuscles are concerned, would be the same as in the previous case. The causes, however, would be different. In this case, too, there would be a tendency to increase of fibrin.

These *six conditions* appear to exhaust all the hypothetical conceptions which are possible in relation to the morbid conditions of the blood found in anæmia. It is of course possible that more conditions than one are being grouped together under this head, and, if this is so, we may hope, by a very rigid analysis, to succeed in differentiating them. At the outset, we notice two forms, which have at least the appearance of being distinct, viz., one in which there is both a diminution of the number of corpuscles and a decrease of hæmoglobin, and a second in which the hæmoglobin alone is deficient, the number of corpuscles remaining normal. The former is by far the most frequent condition, and will, therefore, receive consideration first.

If we take the various examples of anæmic blood, the state of which has been as far as possible accurately made out by chemical analysis, and by corpuscular enumeration and analysis combined, we shall find that we have to deal with a condition in which the tendency is to a steady decrease in number of the red corpuscles, associated with a decrease in the hæmoglobin, altogether out of proportion to the number of corpuscles wanting.

Suppose, for example, the number of corpuscles per cubic mm. to be in the normal state of things 5,000,000, and the average amount of hæmoglobin per corpuscle to be 30 $\mu\mu$ gr.,[*] it is no uncommon case to find the corpuscles reduced in anæmia to such numbers as 2,800,000, and the hæmoglobin to 17·14 $\mu\mu$ gr. per corpuscle for this number. In order to exhibit the matter plainly, we will take round numbers, and consider the corpuscles per cubic mm. to be reduced one-half, viz., to 2,500,000, and the hæmoglobin average per corpuscle to 15 $\mu\mu$ gr. We now see that while the corpuscles per cubic mm. have been reduced to one-half, the hæmoglobin amount per cubic mm. is

---

[*] These numbers are based upon the determination of Malassez and Gowers.

The number of corpuscles is obtained by counting, and the average amount of hæmoglobin by analysis.—*Vide* Malassez's Papers.

"La richesse du sang ne s'évalue pas seulement en comptant le nombre des globules . . . . . il faudrait pouvoir apprécier la quantité d'hémoglobine comprise dans chaque globule."—Académie des Sciences, séance du 2 Décembre, 1872.

Voyez aussi: De la Numération des Globules, 1873, p. 6, et le Mémoire Suivant, p. 31.

" Sur les Diverses Méthodes de Dosage de l'Hemoglobine, et sur un Nouveau Colorimètre. ' Archives de Physiologie,' 1877, p. 1—40."

Gowers's Papers.

Lancet—" On the numeration of blood corpuscles," December 1877, p. 797.

"Practitioner," July, 1878.

" British Medical Journal," May, 1881.

reduced to one-fourth. In other words, a reduction of one-half in the number of corpuscles has involved a reduction of three-fourths in the hæmoglobin. It is quite clear that no explanation can be afforded of a fact like this, so long as we regard the hæmoglobin to be disposed in equal amounts among the corpuscles.

These reductions appear to follow a definite law. The corpuscles are reduced $\frac{2}{4}$th, and the hæmoglobin $\frac{3}{4}$th. The ratio of reduction is, therefore, as 2 to 3.

To understand how such effects as these are brought about it is necessary to study the actual mode of distribution of the hæmoglobin in the corpuscles.

In my first proposition I have stated that the bodies destined to become red corpuscles enter the blood as *colourless discs*, and become coloured whilst circulating in this fluid by a gradual, regular, and continuous *production* of hæmoglobin. From this we may infer the truth that a corpuscle becomes coloured, in such a manner as to acquire in equal times, equal increments of hæmoglobin.

If, then, we bear in mind that the corpuscles become coloured in the blood, not being (as some have supposed) brought to it ready coloured, and that they do not become coloured *suddenly,* but only in a *gradual* manner, it will be obvious that the amount of hæmoglobin which they possess will depend upon the length of time they have been resident in the blood. It is apparent also from these considerations, that, although the hæmoglobin distributed among the 5,000,000 corpuscles belonging to a cubic mm. may give an average of 30 $\mu\mu$ gr. to each, that this by no means indicates its true disposition, for it will exist in some of these corpuscles in every degree below 30, and in others greatly above it—enough above it in fact to enable an average of 30 to be struck.

Suppose, for example, we take (in imagination) increments of hæmoglobin, which are equal in amount to each other, and add them to colourless discs, in such a manner that the first disc shall have one increment only, the second two, the third

three, and so on till we arrive at a number of increments,
which, divided by the number of corpuscles required to be
taken, will yield an average of 30 increments to each corpuscle.
We shall find the number of increments necessary to do this
will be 1,770, and the number of corpuscles in which they are
disposed will be 59.

1,770 increments are, therefore, distributed among every
59 corpuscles, which we will call a *group*. If, then, we divide
the total amount of hæmoglobin, viz., 150,000,000 increments
by 1,770, or the total number of corpuscles (5,000,000) by 59,
we get $84,745\frac{45}{59}$ (*vide* foot-note as to cause of fractions, Table
A, next to page 164), which represent the number of *groups* of
59 corpuscles and *sets* of 1,770 hæmoglobin increments, which
repeat themselves in 5,000,000 corpuscles, and in their hæmoglobin increments this latter number ($84,745\frac{45}{59}$), in contradistinction to *group*, we will call a *set*.

If now we divide up the *time* occupied by a single
corpuscle in attaining its maximum degree of colour, or
amount of hæmoglobin into 59 equal portions, we may say—

1.—A single corpuscle has 59 stages of colouration, and
at the final stage it has in it 59 times as much hæmoglobin as
it possessed at the termination of its first stage.

2.—A *Group* consists of 59 corpuscles, and its individual
corpuscles represent at the same time, or at once, all the stages
which occur in a single corpuscle. The group-figure in any
case will always bear a definite relation to the average number
of hæmoglobin increments per corpuscle. It is, as before said,
the number of corpuscles, which, acting as a divisor to the sum of
the increments, will yield the required average number of increments per corpuscle. In the example before us this average is
obtained by dividing the amount of hæmoglobin per cubic mm.
by the number of corpuscles per cubic mm., and, therefore, the
average number of increments per corpuscle in this case gives
the hæmoglobin value per increment; but this is not so in
every instance. (*Vide* page 175).

3.—When a cubic mm. contains 5,000,000 corpuscles, and
an average of hæmoglobin increments of 30 per corpuscle,

it possesses 59 *sets* of corpuscles, each consisting of 84,745$\frac{4}{59}$ corpuscles. Each *set* represents a distinct stage of colouration ; and all the corpuscles in the same *set* are in the same stage ; *e.g.*, the first *set* has in it 84,745$\frac{4}{59}$ units or increments of hæmoglobin, and the highest, or final set, 59 times this number of units, that is 5,000,000.

It must be borne in mind that the advanced lymph corpuscles or colourless discs leave the blood-glands and enter the blood in *batches* at regular intervals or periods, and as a consequence, the corpuscles of one batch have attained a certain amount of hæmoglobin before they are joined by those of succeeding batches. Now, if we take four hours as the interval between the acts of excitement and turgescence of the blood-glands which accompany the digestive process, we may consider the amount of hæmoglobin which each corpuscle attains during one of these intervals to represent an increment, and in this case, after 60 consecutive meals had been taken, the batch first thrown in would have arrived at its maximum stage of colouration, viz., 59 degrees, and the last batch but one would now represent corpuscles containing only one increment each of hæmoglobin, while the last or 60th batch would consist of colourless discs simply.

All the stages of the development of the red corpuscles would be represented in these batches, and the age of the fully coloured corpuscles would be nearly three weeks—for such a view clearly makes the rate of the colouration of the corpuscles to go on at an average of one increment (1 $\mu\mu$ gr.) in eight hours. If this is so, for a set of colourless discs to obtain 59 increments each, 472 hours, or 19 days 16 hours, will be required. This is, therefore, probably about the time the disc takes to arrive at its maximum colour, if the number of corpuscles in the cubic mm. is 5,000,000, and the average number of hæmoglobin increments 30.*

---

* Everything seems to indicate that the corpuscles continue to form hæmoglobin throughout the whole period of their existence, and that they are destroyed when this power is exhausted.

Having now obtained a better idea of the normal mode of distribution of the hæmoglobin, we are in a position to consider with profit the nature of the changes which result in anæmia.

These will be best understood by reference to Table A, which has been calculated on the basis of the foregoing principles. The effect produced upon the average per corpuscle of the hæmoglobin by reduction of the number of corpuscles can be seen at a glance, and it will be observed that it follows the law of relation of 2 to 3 previously enunciated. The table in fact shows the manner in which the hæmoglobin is arranged in groups of 59 corpuscles, repeated $84,745\frac{45}{59}$ times, when there are 5,000,000 corpuscles in a cubic mm., and an amount of hæmoglobin per cubic mm., which gives an average of 30 μu gr. to each corpuscle, and also illustrates how the decrease of the number of the corpuscles is attended by a *disproportionate* decrease, in the average amount of hæmoglobin per corpuscle.

Consider each number in the *first column* to be included in a similar ring to that at its head $\left(\ 0\ \right)$ which is intended to represent a corpuscle, then the numbers will indicate the number of increments per corpuscle, arranged according to the graduated method. The corpuscles being 59 in number will contain among them 1,770 increments.

The second column shows the number of increments of hæmoglobin for a group of 59, and also for any smaller group than this. Such groups include the group-figure taken, added to those above it; *e.g.*, suppose we take the group-figure 9, then $9+8+7+6+5+4+3+2+1 = 45$ hæmoglobin increments, the amount for group 9, seen opposite to it, and so on with the rest.

The third column gives the *average* number of hæmoglobin increments per corpuscle for each group, and is obtained by dividing the figures in the second column by the group-figure opposite to it in the first; *e.g.*, $\frac{28}{7}=4$, the average number of increments per corpuscle for a group of seven.

The fourth column gives the number of corpuscles per mm. cube for each group-figure; *e.g.*, say we take group-figure 11, the

TABLE A.

			...,... $^{...}/_{59}$	do.	...th	72,9... $^{...}/_{59}$
41	861	21·	3,474,576 $^{16}/_{59}$	do.	41st	72,966,101 $^{41}/_{59}$
42	903	21·5	3,559,322 $^{3}/_{59}$	do.	42nd	76,525,423 $^{44}/_{59}$
43	946	22·	3,644,067 $^{47}/_{59}$	do.	43rd	80,169,491 $^{34}/_{59}$
44	990	22·5	3,728,813 $^{38}/_{59}$	do.	44th	83,898,305 $^{5}/_{59}$
45	1035	23·	3,813,559 $^{19}/_{59}$	do.	45th	87,711,864 $^{24}/_{59}$
46	1081	23·5	3,898,305 $^{5}/_{59}$	do.	46th	91,610,169 $^{9}/_{59}$
47	1128	24·	3,983,050 $^{50}/_{59}$	do.	47th	95,593,220 $^{30}/_{59}$
48	1176	24·5	4,067,796 $^{36}/_{59}$	do.	48th	99,661,016 $^{56}/_{59}$
49	1225	25·	4,152,542 $^{22}/_{59}$	do.	49th	103,813,559 $^{19}/_{59}$
50	1275	25·5	4,237,288 $^{·}/_{59}$	do.	50th	108,050,847 $^{27}/_{59}$
51	1326	26·	4,322,033 $^{53}/_{59}$	do.	51st	112,372,881 $^{21}/_{59}$
52	1378	26·5	4,406,779 $^{40}/_{59}$	do.	52nd	116,779,661 $^{1}/_{59}$
53	1431	27·	4,491,525 $^{25}/_{59}$	do.	53rd	121,271,186 $^{36}/_{59}$
54	1485	27·5	4,576,271 $^{11}/_{59}$	do.	54th	125,847,457 $^{47}/_{59}$
55	1540	28·	4,661,016 $^{56}/_{59}$	do.	55th	130,508,474 $^{14}/_{59}$
56	1596	28·5	4,745,762 $^{42}/_{59}$	do.	56th	135,254,237 $^{17}/_{59}$
57	1653	29·	4,830,508 $^{28}/_{59}$	do.	57th	140,084,745 $^{45}/_{59}$
58	1711	29·5	4,915,254 $^{14}/_{59}$	do.	58th	145,000,000
59	1770	30·	5,000,000	do.	59th	150,000,000

* In 5,000,000 Corpuscles there are 81,745 Groups of 59 Corpuscles, or 59 Sets of 81,745 Corpuscles each. Each of these Sets represents a distinct grade of colour from the minimum or Set 1, in which each Corpuscle contains one increment only to the maximum or Set 59, which contains 59 increments.

† These fractions are simply the result of taking a level number of 5,000,000 Corpuscles per cubic mm. to represent normal blood. If we take the number as 5,000,014, the fractions are avoided.

TABLE A.

This Table is based upon 5,000,000 Corpuscles, and 0·150 Milligrammes of Hæmoglobin per Cubic Millimetre, and 30 μ μ grms. per Corpuscle. Value of Increment, 1 μ μ grms.

	Corpuscle-Group and number of Hæmoglobin Increments in each Corpuscle up to 29.	Increment Group Figure, or the Sum of Hæmoglobin Increments up to 29.	Average Hæmoglobin Increments for each Corpuscle.	Number of Corpuscles per Cubic Millimetre for each Group-Figure up to 29, and of Hæmoglobin Increments per Set for each degree of Colouring.	The Set-frame which is the number of time, with which this degree of Colouring each Corpuscle is equal to 4.	Degree or Grade of Colours and number of Set.	Total Number of Hæmoglobin Increments for the Output belonging to each Group-Figure opposite.
Colourless	0	0	0	†84,745 44/50	0	0	0
	1	1	1·	84,745 44/50	†84,745 44/50	1st	†84,745 44/50
	2	3	1·5	169,491 31/50	do.	2nd	254,237 0/50
	3	6	2·	254,237 0/50	do.	3rd	508,474 21/50
	4	10	2·5	338,983 2/50	do.	4th	847,457 27/50
	5	15	3·	423,728 49/50	do.	5th	1,271,186 26/50
	6	21	3·5	508,474 24/50	do.	6th	1,779,661 1/50
	7	28	4·	593,220 46/50	do.	7th	2,372,881 31/50
	8	36	4·5	677,966 4/50	do.	8th	3,050,847 2/50
	9	45	5·	762,711 44/50	do.	9th	3,813,559 9/50
	10	55	5·5	847,457 7/50	do.	10th	4,661,016 06/50
	11	66	6·	932,203 02/50	do.	11th	5,593,220 3/50
	12	78	6·5	1,016,949 4/50	do.	12th	6,610,169 07/50
	13	91	7·	1,101,694 51/50	do.	13th	7,711,864 27/50
	14	105	7·5	1,186,440 08/50	do.	14th	8,898,305 1/50
	15	120	8·	1,271,186 26/50	do.	15th	10,169,491 01/50
	16	136	8·5	1,355,932 2/50	do.	16th	11,525,423 07/50
	17	153	9·	1,440,677 2/50	do.	17th	12,966,101 03/50
	18	171	9·5	1,525,428 43/50	do.	18th	14,491,525 3/50
	19	190	10·	1,610,169 25/50	do.	19th	16,101,694 44/50
	20	210	10·5	1,694,915 1/50	do.	20th	17,796,610 49/50
	21	231	11·	1,779,661 1/50	do.	21st	19,576,271 31/50
	22	253	11·5	1,864,406 2/50	do.	22nd	21,440,677 3/50
	23	276	12·	1,949,152 2/50	do.	23rd	23,389,830 0/50
	24	300	12·5	2,033,898 26/50	do.	24th	25,423,728 27/50
	25	325	13·	2,118,644 4/50	do.	25th	27,542,372 2/50
	26	351	13·5	2,203,389 42/50	do.	26th	29,745,762 27/50
	27	378	14·	2,288,135 50/50	do.	27th	32,033,898 48/50
	28	406	14·5	2,372,881 31/50	do.	28th	34,406,779 3/50
	29	435	15·	2,457,627 7/50	do.	29th	36,864,406 18/50
	30	465	15·5	2,542,372 2/50	do.	30th	39,406,779 0/50
	31	496	16·	2,627,118 47/50	do.	31st	42,033,898 17/50
	32	528	16·5	2,711,864 24/50	do.	32nd	44,745,762 07/50
	33	561	17·	2,796,610 22/50	do.	33rd	47,542,372 24/50
	34	595	17·5	2,881,355 31/50	do.	34th	50,423,728 06/50
	35	630	18·	2,966,101 03/50	do.	35th	53,389,830 0/50
	36	666	18·5	3,050,847 2/50	do.	36th	56,440,677 2/50
	37	703	19·	3,135,593 12/50	do.	37th	59,576,271 0/50
	38	741	19·5	3,220,338 03/50	do.	38th	62,796,610 3/50
	39	780	20·	3,305,084 44/50	do.	39th	66,101,694 47/50
	40	820	20·5	3,389,830 0/50	do.	40th	69,491,525 05/50
	41	861	21·	3,474,576 24/50	do.	41st	72,966,101 01/50
	42	903	21·5	3,559,322 1/50	do.	42nd	76,525,423 07/50
	43	946	22·	3,644,067 0/50	do.	43rd	80,169,491 42/50
	44	990	22·5	3,728,813 2/50	do.	44th	83,898,305 2/50
	45	1035	23·	3,813,559 9/50	do.	45th	87,711,864 21/50
	46	1081	23·5	3,898,305 1/50	do.	46th	91,610,169 4/50
	47	1128	24·	3,983,050 42/50	do.	47th	95,593,220 27/50
	48	1176	24·5	4,067,796 08/50	do.	48th	99,661,016 09/50
	49	1225	25·	4,152,542 2/50	do.	49th	103,818,559 12/50
	50	1275	25·5	4,237,288 1/50	do.	50th	108,050,847 2/50
	51	1326	26·	4,322,033 07/50	do.	51st	112,372,881 41/50
	52	1378	26·5	4,406,779 0/50	do.	52nd	116,779,661 1/50
	53	1431	27·	4,491,525 04/50	do.	53rd	121,271,186 08/50
	54	1485	27·5	4,576,271 31/50	do.	54th	125,847,457 7/50
	55	1540	28·	4,661,016 06/50	do.	55th	130,508,474 42/50
	56	1596	28·5	4,745,762 42/50	do.	56th	135,251,237 1/50
	57	1653	29·	4,830,508 07/50	do.	57th	140,084,745 2/50
	58	1711	29·5	4,915,254 44/50	do.	58th	145,000,000
	59	1770	30·	5,000,000	do.	59th	150,000,000

* In 5,000,000 Corpuscles there are 84,745 Groups of 59 Corpuscles, or 59 Sets of 84,745 Corpuscles each. Each of these Sets represents a distinct grade of colour (from the minimum or Set 1, in which each Corpuscle contains one Increment only to the maximum or Set 59, which contains 59 Increments.

† These fractions are simply the result of taking a level number of 5,000,000 Corpuscles per cubic mm. to represent normal blood. If we take the number as 3,000,014, the fractions are avoided.

number of corpuscles will be $982,203\frac{23}{59}$ ($11 \times 84,745\frac{45}{59}$), which is also the number of hæmoglobin increments for this set.

The fifth column shows the number of corpuscles in each *set*, all of the corpuscles of which contain the same number of increments of hæmoglobin per corpuscle as the number in the disc opposite to it in the first column, and this number, therefore, represents also the stage or degree of colouration of each *set*.

The sixth column shows the total number of increments of hæmoglobin per cubic mm. for the group it is opposite ; *e.g.*, group 5, contains 15 increments of hæmoglobin, which, multiplied by $84,745\frac{45}{59}$, equals $1,271,186\frac{36}{59}$.

In applying this table to illustrate anæmic conditions we must first ascertain the number of corpuscles per cubic mm. in the specimen of blood ; *e.g.*, suppose such specimen to give 2,881,355 per cubic mm., we shall find the group-figure corresponding to this number to be 34, the number of the hæmoglobin increments in the group to be 595, the average number of hæmoglobin increments per corpuscle 17·5, and the number of hæmoglobin increments per cubic mm. to be $50,423,728\frac{48}{59}$.

The absolute value of the increment, *i.e.*, the hæmoglobin unit, may be seen at the top of the table. In this case 1 $\mu\mu$ gr.

This, of course, proceeds upon the assumption that in anæmic blood the production of hæmoglobin by the corpuscles goes on at the normal rate ; but if this is not the case, then other abnormal conditions, besides the decrease in the number of the corpuscles, are present, and the result will be modified accordingly. The table is, therefore, simply intended to show that *when the mode of distribution of the hæmoglobin is properly understood a decrease in the number of the corpuscles will necessarily involve a disproportionate decrease in the hæmoglobin in the ratio of 2 to 3.*

As before said, the one condition *invariably* present in anæmia is the reduction of the *average* hæmoglobin per corpuscle, and as the hæmoglobin is a *product* of the corpuscles its *average* can only be reduced by the operation of one or other of two things, or of both these acting in concert.

1st.—Either the premature death of the corpuscles

or the increased destruction of the older ones, which amounts practically to the same thing, and the influence of either of which it is the chief purpose of the table to show ; or, 2nd, their inability to produce hæmoglobin at the normal rate, which may arise either from defect in the corpuscles themselves, as before stated, or from a deficiency of the material which they take up and convert into this substance.

Although a deficiency of the average per corpuscle of hæmoglobin pertains to all anæmias (except the hæmorrhagic), they are, nevertheless, capable of division into several forms.

1.—We have the *typical form*, in which the definite law of relation of 2 to 3 obtains between the decrease of the corpuscles and the hæmoglobin. This form, as we have seen. we are able to understand and tabulate.

2.—A form in which the decrease in the hæmoglobin per corpuscle, in relation to the number of corpuscles, is much greater than in the former or typical kind. Take for illustration 3,600,000 to 10·55 $\mu\mu$ gr. of hæmoglobin per corpuscle, instead of about 22 $\mu\mu$ gr. as in the typical kind.

3.—A variety in which, although the average per corpuscle is but a little over half that in health, it is nevertheless higher (in relation to corpuscles) than in the typical kind; *e.g.*, 1,830,000 corpuscles to 18·03 per corpuscle average, instead of about 11·5.

4.—A kind in which the average amount of hæmoglobin per corpuscle diminishes, while the number of corpuscles remains normal, or nearly so.

All these varieties except the fourth conform with each other in one particular, viz., their average hæmoglobin per corpuscle is *much* reduced. The three first agree also in the fact that a diminution of the corpuscles occurs simultaneously with the reduction of the hæmoglobin ; they differ in that this reduction is greater in the one case and less in the other than in the typical variety which follows the definite law. The fourth kind is, however, entirely anomalous, for the corpuscles are present in sufficient numbers ; but they do not produce quite their normal amount of hæmoglobin. It is this capacity of the hæmoglobin to diminish or increase a little without the number of the corpuscles being affected which causes the ordinary anæmias to deviate from the *typical form* in which the hæmoglobin deficiency is entirely accounted for by the corpuscular loss.

The view which best commends itself to one's judgment as to the general cause of anæmia is that set forth in my second hypothesis, viz., that it has its origin in defects of nutrition, arising out of disturbances of the primary digestion. It is easy to conceive that the lymphatic products may receive profound modifications from a cause so capable of altering the nutrient qualities of the liquid of the blood. Such modifications might show themselves in the corpuscles in a variety of ways, all indicating diminished vitality, *e.g.*, (*a*) the life-period of the corpuscles may be diminished without interference with the number produced; (*b*) their function of producing hæmoglobin may be depressed; (*c*) the modifications may be so profound as to produce not only a shortened life but also a diminished proliferation. They may not be, in fact are not, invariably so influenced as to undergo all these modifications at the same time; but it is possible these may sometimes coexist in varying degrees.

The table shows us what would be the effect of early death, or undue destruction of the older and more coloured corpuscles in instances in which the normal numerical supply was kept up and the normal rate of colouration maintained. The group-figure may be taken in any case, as representing the length of the life of the corpuscles, and from this we get at once at the number of corpuscles there should be under these circumstances, per cubic mm., and the average number of increments, and the amount of hæmoglobin per corpuscle; for example, let us suppose that the vitality of the corpuscle is expended by the time it reaches the life-period, represented by 44, then this number will show the mode of grouping of such blood, and opposite this figure in the table we shall find that the number of corpuscles it will yield per cubic mm. will be 3,728,813, and that the average

hæmoglobin per corpuscle will be 22·5. This is an example
of the *typical* variety of which forms 2 and 3 are modifications.

In the second form the interference with the vitality of
the corpuscles has shortened its life to 43, as seen by its
corpuscular number (3,600,000), and has also depressed its
hæmoglobin-forming power, so that instead of yielding an
average of 22, which it should do for this life period, it yields
about half that amount, or 10·55. The ratio of reduction
between the corpuscles and hæmoglobin is as 2 to 6, instead as
in the previous case 2 to 3. The rate of colouration fell, therefore, in this case to nearly half the normal.

The same principles operate, therefore, here as in the
typical cases, reducing the corpuscles and the hæmoglobin
average ; but this average is further reduced, as before
explained, by the interference with the hæmoglobin-
producing power of the corpuscles. In the third variety,
which is said to be one of pernicious or essential anæmia,
the same principles appear to be in operation as in the typical
cases, for the corpuscles are greatly reduced in number
(1,830,000,) so reduced that their group-figure, and, therefore,
there life-period, is as low as 22. The hæmoglobin average for
this number is 11·5, whereas in this case the average per
corpuscle is 18·03, close to the average belonging to the
higher corpuscular number of 2,966,101. The hæmoglobin
average, per corpuscle, is, therefore, 6·53 in excess of what it
should be for the life period represented by 1,830,000 corpuscles,
or by the group-figure 22, and this would go to show that these
corpuscles have been acquiring colour, at a rate more than half
in excess of their normal rate, unless we consider that in such
cases the lowering of the vitality has reached such a point as
not only to bring about a shortened life, but also a diminished
corpuscular supply.

Such a condition would, of course, aggravate the corpuscular
deficiency, and permit us to consider that the corpuscles had
been in the blood for a longer period, in fact, long enough to
obtain an average of 18·03 per corpuscle, *i.e.* to say, they had

succeeded in realising a life-period of 35, a figure still below
the average life-period of ordinary anæmias. This view, of
course, assumes that though diminished in number, the
corpuscles have still a greater life capacity than is indicated by
their number, per cubic mm., that shown by the hæmoglobin
average 18·03, and the numerical deficiency of the corpuscles is
referred to depressed activity of the gland functions.

This latter is probably the true explanation, and the pathological fact which distinguishes pernicious or essential anæmia from
other forms. Looking at it from this point of view, we see that for
an average of 18·03 per corpuscle, there should be 2,966,101
corpuscles per cubic mm.; but as there are only 1,830,000, the
difference between these two figures 1,136,101 will indicate the
extent to which the supply is deficient, which is in round
numbers about one-third.

May not this explain why this form of anæmia has acquired
the qualification pernicious or essential, seeing that we have
to deal not merely with that deficiency of hæmoglobin which
arises from early death of corpuscles, but also with that due
to a deficiency in the initial corpuscular supply. These
conditions, acting in combination, would rob the blood rapidly
of its accumulated capital in the shape of morphological
elements and hæmoglobin, and so destroy by a quick and
double stroke its respiratory relation to the organism.[*]

We come now to consider the fourth variety, in which the
corpuscles are normal, or nearly so, in number, but the average
amount of hæmoglobin per corpuscle is reduced. In such cases
the hæmoglobin could not be materially reduced, without
leaving a larger proportion of discs than usual in the invisible
state, and, therefore, in a condition incapable of being counted.[†]

---

[*] _Vide_ further remarks on this variety, page 171.

[†] In accordance with the views of the blood which I entertain—if
5,000,000 corpuscles existed in a cubic mm., and possessed the colour of
the liquor sanguinis only, they could not be seen, and therefore could
not be counted. Now, suppose such corpuscles to get about half their
normal quantity of hæmoglobin, and this to be distributed in the way
we have shown it must be, would it not follow that the countable

This fact alone would render the countings in such cases unreliable, and when we add to it the further consideration that numbers of the more slightly coloured corpuscles would by the action of the diluting fluid used, and essential to the process, be reduced to invisibility, we are able to see how thoroughly defective this method of counting in this condition would be. In such states, too, the amount of fibrin in the blood would be excessive in the ratio of the lowering of the hæmoglobin. These things considered, it is not probable that the hæmoglobin falls to any considerable degree, while the number of corpuscles remains normal, or nearly so.*

Malassez, in some experiments upon fowls kept under unhealthy conditions, found that the primary effects produced consisted in a slight diminution of the average amount of hæmoglobin per corpuscle without reducing the number of the corpuscles. As the corpuscle is the hæmoglobin producing organ, this would seem to imply that its vitality becomes reduced, and as a consequence it suffers in some instances, 1st, in its hæmoglobin producing function; 2nd, in the length of its life; 3rd, in its numerical productiveness. It is conceivable, too, that corpuscles may be affected in these respects in varying degrees, and this would tend to bring about some of the irregularities which are observed.†

corpuscles would fall very much below 5,000,000? Indeed as the 5,000,000 were rendered visible by the *whole* amount of hæmoglobin, we are justified in supposing that *half* this quantity would fail to make visible about 200,000. In these cases, therefore, in which the normal or nearly the normal number of corpuscles are countable, the hæmoglobin deficiency must be comparatively little.

* The maintenance of the numerical standard of the corpuscles of course excludes any lowering of the hæmoglobin by early death, or anything equivalent to it (*e.g.*, increased ratio of destruction). It can only be attributed to a slower and more imperfect formation of this substance.

† Another observation of Malassez serves to indicate how independent these three powers of the corpuscles may be of each other. In a case of chlorosis, which improved under treatment, while the actual number of corpuscles per cubic mm. diminished, the amount of hæmoglobin per corpuscle became almost double that which it had previously

These three conceptions appear to be competent to explain all the variations which are seen in anæmia from the simplest form, in which the only change is a slight decrease in the amount of hæmoglobin to the ordinary typical form, in which the first and second conditions co-operate to the reduction of both the corpuscles and the hæmoglobin average, to that most grave pathological state *pernicious* anæmia, in which the three conditions may conspire to one common end, viz., to the rapid and entire suppression of the whole of the corpuscular elements of the blood.

It is unfortunate that corpuscle-enumeration cannot be carried out without artificial dilution of the blood, or that this cannot be accomplished by means of liquor sanguinis, or fresh serum, derived by some ready method from the *same* blood, for nothing else can be relied upon to absolutely preserve the red corpuscles from change, and even this does it imperfectly after the blood is shed, as I have shown in those experiments in which I have endeavoured to prevent such changes by strengthening the colloid elements of this fluid.*

When the usual fluids for diluting the blood are used, I have found that a very considerable number of corpuscles become faint, and ultimately indistinguishable, and such, of course, escape being counted. The operation of this would of course, be to give a somewhat smaller number of corpuscles per cubic mm. than there should be (*vide* page 177), and this would result in too large an average of hæmoglobin per corpuscle.

Fortunately, when we understand the true manner of the development of the blood, and the meaning of its various constituents, we are able to compare, to some extent, the results of enumeration with those obtained by ordinary analysis.

---

been. The treatment of anæmia should obviously have reference to several objects: the increase of the vitality, of the length of life, the numerical maintenance of the corpuscles, and of their hæmoglobin producing powers. Such a fact as that pointed out by Malassez serves to show that this must probably be sought in the use of more than one remedy.

* *Vide* Section II.

There is, in fact, scarcely a condition which concerns us in these considerations which cannot be dealt with from the side of analysis ; and as the results in this case are obtained from large quantities of blood, and are not dependent upon quantities or numbers obtained by calculations, in which initial errors are necessarily multiplied, they are probably much more accurate.

In such an analysis the parts of each constituent, in relation to 1,000 parts of blood, are generally given ; and, as the solid elements are all estimated in the *dry* state, it follows that their relations to each other, and to the water which they have yielded, will be got at with great accuracy. The *solid* parts of the blood, which concern us in this connection, are the *fibrin* and the *red corpuscles*. I have elsewhere pointed out that the former may, for all practical purposes, be regarded as representing and as synonymous with the colourless discs of the blood. We have, therefore, simply to compare the relative amounts of the fibrin and the red corpuscles to arrive at the numerical relation which exists between the *colourless* and the *red* disc.

By the method we have used to arrive at the true disposition of the hæmoglobin, in cases where 5,000,000 corpuscles per cubic mm. have been counted, and an average of 30 hæmoglobin increments per corpuscle has been obtained by analysis, we have found that the relation of the *colourless* to the *red* disc is as one to 59, hence, supposing 5,000,000 red corpuscles to exist in a cubic mm., if we would know the actual number of all the discs uncoloured and coloured there are present in such an amount of blood we must add to this number $84,745\frac{4}{5}\frac{5}{9}$, as representing the *colourless discs*, making a total of $5,084,745\frac{4}{5}\frac{5}{9}$.

To ascertain, then, the relative number of colourless and red discs in a case of analysis, we have merely to divide the number of parts of the red corpuscles, by the number of parts of fibrin, and we then get at the group-figure for the red corpuscles. Thus, for example, taking Becqueril and Rodier's tables, we get for the healthy adult man—fibrin, 2·2, corpuscles, 140·0, which gives us 1 colourless disc to 63 red ones.

For the healthy female, in the same table, we get—fibrin, 2·2, corpuscles 127·2 or 1 to $57\frac{9}{11}$.

If, acting on this principle, we take the mean of a number of analyses of healthy, adult male blood, we shall then arrive as closely as it is possible at the exact truth :—

	Fibrin.	Corpuscles.
Becqueril and Rodier	2·2	to 140.
Lehmann	2·025	,, 149·485.
Kirk	2	., 130.
Bennett...........................	3	., 150.
	9·225.	569·485.

This gives us 61 $\frac{8760}{9225}$, which, to avoid fractions, we may regard as 62, and say that the relation of the colourless discs to the red corpuscles in the adult healthy male is 1 to 62.

In this way, we get also the group-number, and by dividing by it, the number of increments, which such a number represents, viz., 1,953, we get the average increments of hæmoglobin per corpuscle, as 31·5. If we retain the same value for the increment of the hæmoglobin (1 $\mu\mu$ gr.) and raise the group-figure, we necessarily increase the average per corpuscle, and therefore its hæmoglobin richness; we also proportionately raise the increment group-number—*e.g.*, suppose on the basis furnished by analysis we raise the corpuscular group-figure to 62, we are compelled to raise also the increment group-number to 1,953, because the first corpuscle in our graduated series must contain one unit of hæmoglobin, and our last 62. This necessarily raises the average number of increments per corpuscle to 31·5, because this average is determined by dividing 1,953 by 62. This rearrangement does not necessarily affect the number of corpuscles per cubic millimetre, which may remain, as before, at 5,000,000, but it does influence the set-figure, because 5,000,000 divided by 62 equals 80,645$\frac{10}{62}$. We also raise the total number of increments of hæmoglobin per cubic millimetre from 150,000,000 to 157,000,000, and this number, divided by 1,953, also yields the set-figure 80,645$\frac{815}{1953}$. This shows that if we raise the group-figure, but retain the same number of corpuscles per cubic millimetre, and the same value for the hæmoglobin increment, we must increase the *amount* of hæmoglobin per cubic millimetre. The amount of this increase in this case is represented by 157,500,000 as against 150,000,000 increments.

If, on the other hand, we desire to raise the group-figure, and to retain the same value for the increment and the same amount of hæmoglobin per cubic millimetre, we must reduce the number of corpuscles from 5,000,000 to 4,761,904$\frac{340}{313}$ per cubic millimetre.

Again, if we desire to raise the group-figure, but to retain at the same time the number of corpuscles at 5,000,000 and to retain the same amount of hæmoglobin per cubic millimetre, and therefore the same average *amount* per corpuscle, we can only do this by creating a difference between the value of the individual increments arrived at by dividing 1,953 by 62, and 150,000,000 by 5,000,000, the former giving us 31·5 and the latter 30 as the *number* of increments per corpuscle. The *value* of the increment in the first case would, therefore, become less than 1 $\mu\mu$ gr., as much less as the difference between 30 and 31·5.

The raising of the group-figure, therefore, compels us either to increase the *number* of hæmoglobin increments per cubic millimetre or to lessen their *value*.

If now, we view adult female blood in the same light, we find that no reasons exist why we should regard the group-figure, or the average hæmoglobin increments per corpuscle, as different to the male,—nor can we allow that in a condition of equilibrium of female blood there would be more than one colourless disc to the group-figure. Unfortunately we do not possess a sufficient number of analyses of female blood to enable us to settle this question by taking the mean of a number, as we have done with the male, but from the few examples we have, we are able to see that the group-figure 62 could easily be obtained by regarding the relation of fibrin and corpuscles, as 2·05 to 127, figures quite accessible and legitimate.

Thus, the difference between male and female blood is made to consist in a variation of the number of corpuscles per cubic mm., and this is determined by the number of times

Table B.—Female.*

This Table is based upon 4,500,000 Corpuscles, and 0·135 Milligrammes of Hœmoglobin per Cubic Millimetre, and 3·45 μ μ grms. per Corpuscle.   Value of Increment, 1 μ μ grms.

	Corpuscle-order (or up-figure) and Number of Increments to each Corpuscle.	Incremental Group-Figure of Hœmoglobin-values of Increments in each Group-Figure.	Average Hœmoglobin-Increment per Corpuscle in each Group-Figure.	Number of Corpuscles per Cubic Millimetre for each Group-Figure up to and at Hœmoglobin-Increment per Corpuscle for each degree of colouration.	The set-figure which, with number, at present, or if it make up a previous set-figure = opposite Corpuscle.	Decolouration or tint of Corpuscle, and Number of set.	Total Number of the Hœmoglobin-Increments for being to be a Colour-Figure opp. to.
Colourless	0	0	0	72,580⁰	0	0	72,580 ⁰/₃
	1	1	1·	72,580⁰/₄₅	72,580⁰/₄₅	1st	72,580 ⁰/₃
	2	3	1·5	145,161 ⁰/₄	do.	2nd	217,711⁰/₃₁
	3	6	2·	217,741²/₄₁	do.	3rd	435,483²/₃₁
	4	10	2·5	290,322⁴/₄	do.	4th	725,806⁴/₃
	5	15	8·	362,903 ⁷/₄₅	do.	5th	1,088,709⁴/₃₁
	6	21	3·5	435,483²/₄	do.	6th	1,524,193⁶/₄
Limit of essential or pernicious Anæmia.	7	28	4·	508,064¹⁸/₄	do.	7th	2,032,258 ²/₄
	8	36	4·5	580,645 ⁷/₃₁	do.	8th	2,612,903 ⁷/₄
	9	45	5·	653,225²/₄	do.	9th	3,266,129 ¹/₄
	10	55	5·5	725,806⁴/₄	do.	10th	3,991,935⁴/₄₁
	11	66	6·	798,387 ¹/₄₁	do.	11th	1,790,322⁶/₄₁
	12	78	6·5	870,968⁰/₄	do.	12th	5,661,290⁰/₃₁
	13	91	7·	940,548⁰/₄₀	do.	13th	6,601,838⁰/₄₁
	14	105	7·5	1,016,129 ¹/₄₀	do.	14th	7,620,967²/₄
	15	120	8·	1,088,709⁶¹/₄	do.	15th	8,709,677⁰/₄₁
	16	136	8·5	1,161,290⁰/₄	do.	16th	9,870,967⁰/₄
	17	153	9·	1,233,870⁰/₄	do.	17th	11,104,838⁴/₄₁
	18	171	9·5	1,306,451⁰/₄	do.	18th	12,111,290⁰/₃₁
	19	190	10·	1,379,032 ⁷/₄	do.	19th	13,790,322⁰/₄
	20	210	10·5	1,451,612⁰/₄	do.	20th	15,241,935¹⁴/₄
	21	231	11·	1,524,193¹/₄	do.	21st	16,766,129 ¹/₄
	22	253	11·5	1,596,774 ⁰/₃₁	do.	22nd	18,362,903 ⁷/₄
	23	276	12·	1,669,354⁰/₄₀	do.	23rd	20,032,258 ²/₄
	24	300	12·5	1,741,935⁴/₄	do.	24th	21,774,193⁰/₄₀
	25	325	13·	1,814,516 ⁴/₄	do.	25th	23,588,709⁰/₄
	26	351	13·5	1,887,096⁰/₄	do.	26th	25,475,806⁰/₄
	27	378	14·	1,959,677⁰/₄	do.	27th	27,435,483⁰/₄
	28	406	14·5	2,032,258 ²/₄	do.	28th	29,467,741²/₄₁
	29	185	15·	2,104,838⁰/₄	do.	29th	31,572,580⁰/₄₁
	30	465	15·5	2,177,419¹¹/₄	do.	30th	33,750,000
Anæmic region.	31	496	16·	2,250,000	do.	31st	36,000,000
	32	528	16·5	2,322,581⁰⁰/₄	do.	32nd	38,322,580⁰⁰/₄
	33	561	17·	2,395,161 ¹/₄	do.	33rd	40,717,741¹/₄
	34	595	17·5	2,467,741²/₄₁	do.	34th	43,185,483⁰/₄
	35	630	18·	2,540,322⁰/₄	do.	35th	45,725,806¹¹/₄
	36	666	18·5	2,612,903 ⁷/₄	do.	36th	48,338,709⁰/₄
	37	703	19·	2,685,183⁰/₄	do.	37th	51,024,193¹¹/₄
	38	741	19·5	2,758,064⁰/₄	do.	38th	53,782,258 ⁷/₄
	39	780	20·	2,830,645 ⁵/₄₀	do.	39th	56,612,903 ¹/₄
	40	820	20·5	2,903,225²/₄	do.	40th	59,516,129 ⁰/₃₁
	41	861	21·	2,975,806¹¹/₄	do.	41st	62,491,935⁴/₄
	42	903	21·5	3,048,387 ¹/₄	do.	42nd	65,540,322⁰/₄
	43	946	22·	3,120,967²/₄	do.	43rd	68,661,290⁰/₄
	44	990	22·5	3,193,548⁰/₄	do.	44th	71,854,838⁰/₄
	45	1035	23·	3,266,129 ¹/₄	do.	45th	75,120,967⁰/₄
	46	1081	23·5	3,338,709⁰/₄	do.	46th	78,459,677⁰/₄
	47	1128	24·	3,411,290⁰/₄	do.	47th	81,870,967⁴/₄
	48	1176	24·5	3,483,870²/₄	do.	48th	85,354,838²/₄
	49	1225	25·	3,556,451⁰/₄	do.	49th	88,911,290⁰/₄
	50	1275	25·5	3,629,032 ⁷/₄	do.	50th	92,540,322⁷/₄
	51	1326	26·	3,701,612⁰/₄	do.	51st	96,241,935¹/₄
	52	1378	26·5	3,774,193¹¹/₄	do.	52nd	100,016,129 ¹/₄
	53	1431	27·	3,846,774 ⁴/₄	do.	53rd	103,862,903 ⁷/₄
	54	1485	27·5	3,919,354⁰/₄	do.	54th	107,782,258 ⁷/₄
	55	1540	28·	3,991,935⁴/₄	do.	55th	111,774,193⁰/₄
	56	1596	28·5	4,064,516 ¹/₄	do.	56th	115,888,709⁰/₄
	57	1653	29·	4,137,096¹¹/₄₁	do.	57th	119,975,806¹¹/₄
	58	1711	29·5	4,209,677⁰/₄₁	do.	58th	124,185,483⁰/₄
Health range.	59	1770	30·	4,282,258 ⁷/₄	do.	59th	128,467,741⁷/₄
	60	1830	30·5	4,354,838⁰/₄	do.	60th	132,822,580⁰/₄
	61	1891	31·	4,427,419¹¹/₄₁	do.	61st	137,250,000
	62	1953	31·5	do.	do.	62nd	141,750,000

* The Male Table corresponding to this would have a Group-Figure of 62 5,000,000 Corpuscles, and 0·137 Milligramme to the Cubic Millimetre ; a Set Figure of ⁶⁰⁶¹⁵, an Increment value of 1 μμ grms., and a total Incremental Number of 157,500,000.

† In 4,500,000 Corpuscles there are 72,580 groups of 62 Corpuscles, or 63 Sets of 72,580 Corpuscles each. Each of these Sets represents a distinct grade of colour from the minimum or Set, 1, in which each Corpuscle contains one Increment only, to the maximum or Set 62, which contains 62 Increments.

‡ Vide note referring to cause of fractions, Table A.

the groups repeat themselves in 4,500,000 corpuscles, assuming this to be the correct counting for female blood.

It will be noted that the number of increments per corpuscle is based upon the *graduated manner* in which the hæmoglobin is distributed among the corpuscles, and is not got at by dividing the actual amount of hæmoglobin by the number of corpuscles per cubic mm., as in the enumeration methods. Between the group-number decided upon and the average number of increments per corpuscle there is an invariable and definite relation.

The absolute *value* of the increment can only be known in these latter cases by ascertaining the total number of increments in a cubic mm. of blood by multiplying the increment group-number by the set-figure and dividing the total amount of hæmoglobin in a cubic mm. by it.

After we know the value per increment in hæmoglobin weight, we can assign the hæmoglobin value for each individual corpuscle in the group series by multiplying that unit by the group-number.

In Table B will be found all that pertains to a group-figure of 62, and an increment average of 31·5, and a corpuscular number of 4,500,000 per cubic mm. The actual increment value in hæmoglobin weight is given at the head of the table.

It will be well to remember that neither the group-figure, nor the increment average, when based upon the gross analysis of blood, is interfered with as in the enumeration methods by anything which destroys corpuscles or renders them incapable of being enumerated, such as dilution. Such actions reduce the *sets* that should be seen, and render the results to that extent fallacious.

Of course the diluting fluids act upon all the corpuscles alike, but those which have the least colour are the first to disappear, both by the loss of the hæmoglobin they contain and also by the colouring and levelling up of the tint of the liquid which surrounds them.*

Vide Section II.

It will be observed that the method of analysis gives us a slightly higher group-figure and increment-average than we obtain by the method of counting. This elevation of these numbers does not affect the number of corpuscles per cubic mm., nor the actual amount of hæmoglobin per corpuscle. Both these depend upon the number of group-sets, or repetitions of the groups upon which the actual number of corpuscles per cubic mm. in its turn depends. The elevation of the group-number, however, does increase a little the estimation of the life-term of the corpuscle, making it three weeks, instead of as by counting 19 days and 20 hours.

It is most desirable that the enumeration methods should be perfected, as I have before suggested, by the use of fresh serum, to which colloids have been added, for there is no other way besides counting of getting at the *number* of corpuscles per cubic mm.

It is proposed now to illustrate the principles laid down in this paper by an analysis of cases of anæmia, the facts of which have been gathered by independent and competent observers. Malassez, in his paper " Sur la Richesse en Hémoglobine des globules rouges du Sang," furnishes us with twelve cases of anæmia of various kinds, in which the number of corpuscles and the amount of hæmoglobin were carefully estimated. To Malassez's table recording these observations I have appended the deductions which flow from the considerations set forth in Table A. The figures up to the double line belong to Malassez, beyond it they are my own, and to the cases requiring comment I have added notes to which the figures in brackets in the last column refer.

(1) Case II.—The treatment in this case has resulted in elevating the life term of the corpuscles from 34 to 39, and has thus raised the number of corpuscles to 3,300,000, while the hæmoglobin production still lags a little more behind than it did at first. The ratio is, therefore, not quite that of the law of 2 to 3, but nearly so.

(2) Case III.—In this case the treatment has been

APPLICATION OF TABLE A TO THE INTERPRETATION OF M. MALASSEZ'S CASES OF ANÆMIA.

No. of Case and Sex.	Names of the Diseases.	Per Cubic Millimetre. Corpuscles	Per Cubic Millimetre. Hæmoglobin	Number of Corpuscles per Millimetre of Hæmoglobin.	Quantity of Hæmoglobin per Corpuscle. μμ gr.	Group Figure or Life-term of Corpuscle.	True average of Hæmoglobin per Corpuscle for its Life-term or Group Figure.	Deficiency or Excess of Hæmoglobin per Corpuscle.	Kind of Anæmia.
1. Woman ..	Intense Chlorosis at the commencement of treatment	3,600,000	0·038	94,620,000	10·55	43	22·0	−11·45	2nd
2. Woman ..	Chlorotic Anæmia treatment begun January 15th, 1877	2,800,000	0·048	58,330,000	17·14	34	17·5	+ 0·36	1st
„ ..	Improved conditions, February 20th, 1877 ..	3,300,000	0·062	53,220,000	18·78	39	20·0	+ 1·22	„ (1)*
3. Woman ..	Chlorotic Anæmia in course of treatment ..	2,720,000	0·053	51,320,000	19·48	32	16·5	+ 2·98	1st (2)*
4. Woman ..	Hysterical Chlorotic Anæmia	2,720,000	0·082	48,530,000	20·50	48	24·0	+ 3·50	2nd
5. Man ..	Anæmia before any treatment	4,000,000	0·086	46,510,000	21·50	48	24·0	− 3·0	2nd
6. Woman ..	Old Chlorosis treated for several years with Iron	4,000,000	0·091	43,950,000	22·75	48	24·0	− 1·25	2nd
7. Woman ..	General Rheumatism partially cured ..	4,000,000	0·101	39,600,000	25·25	48	24·0	− 1·25	1st
8. Woman ..	Phthisis, third stage	3,480,000	0·058	60,000,000	16·66	41	21·0	+ 4·34	2nd
9. Woman ..	Said to be Essential Anæmia	1,830,000	0·033	55,450,000	18·03	22	11·5	+ 6·53	? (3)*
10. Man ..	Lymphadenoma of the testicle recurring after operation, and becoming general ..	3,200,000	0·077	41,550,000	24·06	38	19·5	+ 4·56	1st (4)*
11. Man ..	Lymphadenoma of the skin without ulceration, December 10th, 1876.. ..	4,700,000	0·101	46,530,000	21·49	56	28·5	− 7·01	2nd
„ ..	26th January, 1877	4,500,000	0·110	40,900,000	24·44	54	27·5	− 3·06	„ (5)*
12. Man ..	Cancer of the stomach, 29th September, 1876, before any treatment	1,960,000	0·024	81,660,000	12·24	23	12·0	+ 0·24	1st
„ ..	6th October, 1876, Iodide of Iron for eight days	2,040,000	0·033	47,440,000	16·17	24	12·5	+ 3·67	1st
„ ..	20th October, 1876, Iodide replaced by Perchloride	2,200,000	0·043	51,160,000	19·54	26	13·5	+ 6·04	1st
„ ..	12th December, 1876, Iron left off for one month	2,000,000	0·026	76,920,000	13·00	24	12·5	+ 0·5	1st
„ ..	19th January, 1877, Milk Diet without Iron ..	1,520,000	0·029	52,440,000	19·07	18	9·5	+ 9·57	1st
„ ..	16th February, 1877, Perchloride of Iron for three weeks	2,140,000	0·043	49,760,000	20·09	25	13·0	+ 7·0	1st (6)*

* The figures to which asterisks are appended refer to notes in the text.   ‡ The figures up to the double line are Malassez's, beyond it my own.

slightly more favourable to the development of hæmoglobin than to increase in corpuscles, consequently the average hæmoglobin amount per corpuscle is 2·98 in excess of the normal for the group-figure 32.

(3) Case IX.—One is not in a position to properly interpret this case, without knowing in the first place whether it had been under treatment before the estimations were made. It is scarcely likely that a case of such gravity, in which the respiratory power of the blood would be so profoundly impaired, would remain without treatment. It is quite possible to view it as a case in which the life-period of the corpuscles has been reduced to 22, but in which the hæmoglobin producing power of the corpuscles has been stimulated by treatment to be equivalent to a life-period of 35. (*Vide* page 168.)

In the event of no treatment having been employed, we may regard the case as one in which the corpuscular supply is deficient, but in which the corpuscles live long enough to obtain an amount of hæmoglobin, capable of giving an average per corpuscle of 18·03, equal to a life-time of 35.

(4) Case X.—This is probably a case of the first variety, in which, owing to the treatment, the hæmoglobin average per corpuscle has been raised 4·56 above the normal for this group-figure or life-period.

(5) Case XI.—This belongs to the second variety, in which the hæmoglobin is much below normal for the life-period indicated by the number of corpuscles. Under treatment the corpuscles fall a little, but the average hæmoglobin per corpuscle rises to 24·44 as against 21·49, so that the normal relation for the group-figure is more approached. The increased fall in the corpuscles is probably due to the progress of the disease, while the rise in the hæmoglobin is due to treatment.

(6) Case XII.—This case is very interesting and instructive, and the more so from the security afforded by the number of observations which Malassez made with it. The

estimations made before any treatment had been adopted at once show it to be a case of the first variety; and on reference to Table A, its group-figure will be found to be 23, and all its figures will be seen to approximate with sufficient closeness to show that it belongs to the *typical* variety. Under treatment with iron the corpuscles and hæmoglobin are both increased, but the latter, as most frequently happens, in excess, so that, after treatment from the 6th of October to the 12th of December, it had made an excess of 6·04 per corpuscle above the normal for the group-figure 26, which it then had.

The iron being left off for a month the case fell back to nearly the condition it had at the commencement, and corresponded closely with the group figure 24, in Table A. On the 19th January estimations were made. In the meantime the patient had been fed upon a milk diet, but iron had been withheld. Under this treatment the corpuscles fell greatly; but the hæmoglobin rose from 0·026 to 0·029, which gave for this small number of corpuscles the large average per corpuscle of 19·07 instead of the normal 9·5 for this group figure (18). Each corpuscle in fact had generated 9·57 more hæmoglobin than normal for its life-period. The excess is only in relation to time, for every one of these corpuscles, if it lived out its full period, would generate 59 such increments of hæmoglobin. The effect is, therefore, to be looked upon as a sort of fillip to the hæmoglobin-producing power of the corpuscles.

In marked contrast was the subsequent effect of the perchloride of iron, which by improving the life term of the corpuscles to 25, raised both the number of corpuscles and the amount of hæmoglobin, and while slightly increasing the average also produced nearly the same degree of excess action for the production of hæmoglobin, on the part of the individual corpuscles, as previously existed.

This patient died early in June after repeated attacks of hæmatemesis. At the autopsy cancer of the stomach was found, which had been latent during life, and had rendered the diagnosis for a long time uncertain.

If we exclude case IX. the considerations which arise out of the table appear to indicate the existence of two varieties of anæmia. 1. One in which the relation of the corpuscles and hæmoglobin fall precisely as they should do in the view which refers anæmia to an abbreviation of the normal life-term of the corpuscles. 2. One in which in addition to this the hæmoglobin-producing function of the corpuscle has become impaired. Both these varieties are amenable to treatment with iron, and in both its exhibition leads to an increase of both corpuscles and hæmoglobin; in other words, it is a remedy which tends to lengthen the life of the corpuscle, and also to improve its hæmoglobin-producing function. It will be interesting to know whether these two operations are simultaneous, or whether one precedes the other in point of time.*

Some investigators (among whom is Dr. Gowers, who has made most valuable improvements in the hæmacytometer) prefer to estimate the corpuscles and hæmoglobin in percentage amounts instead of by the number and quantity per cubic millimetre. The following is Dr. Gowers's method of procedure. 995 cubic millimetres of diluting fluid are added to five cubic millimetres of blood. The counting cell is $\frac{1}{5}$ of a millimetre deep, and the floor of it is ruled into tenth of a millimetre squares. The number in ten squares being counted and multiplied by 10,000 gives the number of corpuscles in a cubic millimetre of blood *i.e.* to say 5,000,000. Taking this number as the average per cubic millimetre the average number in two squares of the cell will be 100. The number in two squares, therefore, expresses the percentage proportion of the corpuscles to that of health, or made into a two place decimal the propor-

---

* As a general rule, the effect of treatment is to bring about an increase in the hæmoglobin average per corpuscle, and this is followed by an increase in the number of the corpuscles, probably owing to the fact that the general nutrition undergoes improvement.

tion which the corpuscular richness of the blood examined bears to healthy blood taken as unity, or the number in 20 squares may be taken as a three place decimal. For instance, if 10 squares contain 355 corpuscles equal to 3,350,000 per cubic millimetre of blood, the average of two squares (71) is the percentage, and thus the corpuscular richness is 0·71. So much for the corpuscles.

The amount of hæmoglobin in a cubic millimetre of blood may be ascertained by the quantitative estimate of the iron it contains, or by the capacity of the hæmoglobin for oxygen; but these methods are too difficult for clinical purposes, and this has led to a plan being devised by which approximately correct results can be arrived at by establishing a definite relation between a standard tint of picro-carminate of ammonia in glycerine jelly and a known dilution of normal blood or of a known amount by weight of hæmo-globin in solution. In the former case the scale is arranged for percentages and in the latter case for fractions of milligrammes. Any of such methods have of course the disadvantages which pertain to visual estimations, and are to some extent related to the capacity which the observer may possess of detecting delicate differences of tint, therefore every person using these instruments will require a certain amount of training before his results will become reliable. Fortunately each observer may test his own capacity by repeated estimations of the same blood. Those who adopt the percentage method may throw their estimations into the groupal form by certain simple calculations by bearing in mind the following :—100 per cent. of corpuscles = 5,000,000 corpuscles per cubic millimetre ; 100 per cent. of hæmoglobin = 150,000,000 increments of hæmo-globin ; $\frac{150000000}{5000000}$ = 30, the average number of increments of hæmoglobin per corpuscle ; an average of 30 increments, arranged in the graduated manner take up 59 corpuscles and 1,770 increments of hæmoglobin—$\frac{5000000}{59}$ and $\frac{150000000}{1770}$ give 84,745$\frac{5}{9}$, the set-figure which shows the number of groups of 59 corpuscles there are in 5,000,000, and also the number of corpuscles of each degree of colouration from the minimum of one increment to the maximum of 59 increments there are in 5,000,000, or in a cubic millimetre of blood. To get then the number of corpuscles in any percentage, 100 = 5,000,000, let x stand for percentage number, then as 100 : x : : 5,000,000 ; *e.g.*, as 100 : 60 : : 5,000,000 is to 3,000,000, the corpuscular number for 60 per cent.

To get the per centage of the hæmoglobin in equal increments 100 = 150,000,000. Let x stand for percentage amount of hæmoglobin, as 100 : x : : 150,000,000, *e.g.* as 100 : 30 : : 150,000,000 : 45,000,000. To find the average number of increments per corpuscle we divide the number of hæmoglobin increments by the number of corpuscles. Thus $\frac{150000000}{5000000}$ = 30 or $\frac{45000000}{3000000}$ = 15. Then to find the group-figure which corresponds to any percentage number we get first the per centage number of corpuscles to the group unit by dividing 100 by 59 = $1\frac{41}{59}$. As this number is to 1 so is any other percentage number to its group-figure. Suppose the percentage number to be 60, then as $1\frac{41}{59}$: 1 : : 60 : 35·4. In Table C the percentage figures which correspond to the group-figures of Table A have for convenience been given in a tabulated form.

TABLE C.

Percentage Number corresponding to each Group-Figure.

Percentage Number.	Group-Figure.	Percentage Number.	Group-Figure.	Percentage Number.	Group-Figure.	Percentage Number.	Group-Figure.
1·69*	1	27·11	16	52·54	31	77·96	46
3·38	2	28·81	17	54·23	32	79·66	47
5·08	3	30·50	18	55·93	33	81·35	48
6·77	4	32·20	19	57·62	34	83·05	49
8·47	5	33·89	20	59·32	35	84·74	50
9·16	6	35·59	21	61·01	36	86·44	51
11·86	7	37·28	22	62·71	37	88·13	52
13·55	8	38·98	23	64·40	38	89·83	53
15·25	9	40·67	24	66·10	39	91·52	54
16·93	10	42·37	25	67·79	40	93·22	55
18·64	11	44·06	26	69·49	41	94·91	56
20·33	12	45·76	27	71·18	42	96·61	57
22·03	13	47·45	28	72·88	43	98·30	58
23·72	14	49·15	29	74·57	44	100·0	59
25·42	15	50·84	30	76·27	45		

* The decimals are carried out to two places only.

By means of these rules the Tables C and D have been constructed to render it easy to compare the percentage numbers of the corpuscles and of the hæmoglobin in any case of anæmia, with the groupal mode of distribution of the hæmoglobin among the corpuscles. Say, for instance, we get a case in which the corpuscles stand at 60 and the hæmoglobin at 30 per cent., we refer to 60 in the percentage column of Table D for the number of corpuscles, and to 30 in the same column for the number of hæmoglobin increments. The latter number divided by the former will give the average number of increments per corpuscle. By referring to Table A, and ascertaining the nearest corpuscular number which corresponds to 60 per cent., we arrive at the group-figure, and are then able to compare the corpuscular average with that obtained by the percentage method. This average may agree with or be above or below the *typical or normal* for the number of corpuscles. If above, the patient is under influences which are favourable to the formation of hæmoglobin (say iron treatment). If below, the hæmoglobin percentage has become impaired by the lowering of the vitality of the corpuscles. On the other hand, if the hæmoglobin averages coincide, it indicates that the life-term alone of the corpuscles is affected, and that they have retained their normal hæmoglobin-producing power unimpaired.

Showing the Number of Corpuscles and Hæmoglobin Increments which
correspond to each Percentage Number from 1 to 100 on the basis
of 5,000,000 Corpuscles and 0·150 Milligrammes of Hæmoglobin
per Cubic Millimetre, and an average of 30 μ μ grms. per Corpuscle.

Per centum.	Corpuscles.	Hæmoglobin Increments.	Per centum.	Corpuscles.	Hæmoglobin Increments.
1	50,000	1,500,000	51	2,550,000	76,500,000
2	100,000	3,000,000	52	2,600,000	78,000,000
3	150,000	4,500,000	53	2,650,000	79,500,000
4	200,000	6,000,000	54	2,700,000	81,000,000
5	250,000	7,500,000	55	2,750,000	82,500,000
6	300,000	9,000,000	56	2,800,000	84,000,000
7	350,000	10,500,000	57	2,850,000	85,500,000
8	400,000	12,000,000	58	2,900,000	87,000,000
9	450,000	13,500,000	59	2,950,000	88,500,000
10	500,000	15,000,000	60	3,000,000	90,000,000
11	550,000	16,500,000	61	3,050,000	91,500,000
12	600,000	18,000,000	62	3,100,000	93,000,000
13	650,000	19,500,000	63	3,150,000	94,500,000
14	700,000	21,000,000	64	3,200,000	96,000,000
15	750,000	22,500,000	65	3,250,000	97,500,000
16	800,000	24,000,000	66	3,300,000	99,000,000
17	850,000	25,500,000	67	3,350,000	100,500,000
18	900,000	27,000,000	68	3,400,000	102,000,000
19	950,000	28,500,000	69	3,450,000	103,500,000
20	1,000,000	30,000,000	70	3,500,000	105,000,000
21	1,050,000	31,500,000	71	3,550,000	106,500,000
22	1,100,000	33,000,000	72	3,600,000	108,000,000
23	1,150,000	34,500,000	73	3,650,000	109,500,000
24	1,200,000	36,000,000	74	3,700,000	111,000,000
25	1,250,000	37,500,000	75	3,750,000	112,500,000
26	1,300,000	39,000,000	76	3,800,000	114,000,000
27	1,350,000	40,500,000	77	3,850,000	115,500,000
28	1,400,000	42,000,000	78	3,900,000	117,000,000
29	1,450,000	43,500,000	79	3,950,000	118,500,000
30	1,500,000	45,000,000	80	4,000,000	120,000,000
31	1,550,000	46,500,000	81	4,050,000	121,500,000
32	1,600,000	48,000,000	82	4,100,000	123,000,000
33	1,650,000	49,500,000	83	4,150,000	124,500,000
34	1,700,000	51,000,000	84	4,200,000	126,000,000
35	1,750,000	52,500,000	85	4,250,000	127,500,000
36	1,800,000	54,000,000	86	4,300,000	129,000,000
37	1,850,000	55,500,000	87	4,350,000	130,500,000
38	1,900,000	57,000,000	88	4,400,000	132,000,000
39	1,950,000	58,500,000	89	4,450,000	133,500,000
40	2,000,000	60,000,000	90	4,500,000	135,000,000
41	2,050,000	61,500,000	91	4,550,000	136,500,000
42	2,100,000	63,000,000	92	4,600,000	138,000,000
43	2,150,000	64,500,000	93	4,650,000	139,500,000
44	2,200,000	66,000,000	94	4,700,000	141,000,000
45	2,250,000	67,500,000	95	4,750,000	142,500,000
46	2,300,000	69,000,000	96	4,800,000	144,000,000
47	2,350,000	70,500,000	97	4,850,000	145,500,000
48	2,400,000	72,000,000	98	4,900,000	147,000,000
49	2,450,000	73,500,000	99	4,950,000	148,500,000
50	2,500,000	75,000,000	100	5,000,000	150,000,000

On the Relation which the Products of the Bone-Marrow and of the Spleen of the Adult Ovipara bear to its Blood.

ONE of the modes of the development of the blood in ovipara is from a remarkably delicate *nucleated elliptical cell.* The exterior of this cell is so colourless and free from granulous appearance as to make it most difficult to be seen, hence, when the blood is examined, numbers of what appear to be free nuclei are alone observed ; but when the blood corpuscles are brought closely together into a single layer, as in the " method of packing," the transparent colourless margins of these cells are rendered obvious, and we then find that an orderly sequence as regards colour prevails between the ordinary red nucleated corpuscles and the clear bordered, colourless, nucleated ellipsoids, which in reality represent the earliest stage in the blood of the red nucleated corpuscle of the ovipara.* The corpuscles here referred to are depicted in Photographs 83 and 84, Plate XIII., which is a specimen from the blood of the perch. The same results are obtainable from the blood of birds and reptiles. (*Vide* also descriptions of Photographs 193, 194, and 195, Plate XXIII.)

In the adult ovipara these corpuscles appear to be produced in the bone-marrow and in the spleen, and to be transmitted thence to the blood in the advanced state in which we find them as the colourless nucleated ellipsoids. It is very difficult to trace

---

* These *colourless nucleated ellipsoids,* the nuclei of which are alone visible, are the *analogue* of the *advanced* lymph, splenic, and bone-marrow copuscles, and therefore of the invisible colourless discs of the mammal.

these bodies in the bone-marrow and the spleen, owing to the
excessive delicacy of the exterior margin and the readiness with
which it becomes dissipated, and this difficulty is further
increased by the tendency which the nucleus has to swell
up under the least disturbing influence.   The delicacy of
these exterior margins, and their tendency to disappear as
granular films upon the surface of the slide, may be judged of
by reference to Photograph 173, Plate XXII., which is a specimen
obtained from the bone-marrow of the common fowl. In Photograph 174, Plate XXII., they have been more perfectly preserved
by the use of saturated solution of osmic acid, and their outlines
having become granulated and stained are distinctly seen. Both
these specimens are from the bone-marrow of the fowl, and
have been in each case stained with fuschine.   I have before
referred to the fact that the nuclei of these clear cells often
swell greatly, and in doing so they encroach upon the clear
margin of the cell, and in some cases appear to cause its
disappearance by entirely filling up its cavity.   Photograph
175, Plate XXII., shows this action in progress, but it is most
complete in Photograph 176, Plate XXII.   These nucleated
ellipsoids of the bone-marrow are in their turn yielded by
pigmentary cells, which may be seen in both the smooth and the
coarsely granular state, and which differ in no essential particular from similar cells of the mammal bone-marrow.   Some of
these parent cells may be seen in Photograph 174, Plate XXII.,
and in Photograph 177, Plate XXII.   The granular *débris* on the
right hand of this specimen is formed by fusion of some of these
cells.   The general analogy of the earliest products of the
oviparous bone-marrow to those of the mammal may be judged
of by Photograph 178, Plate XXII.

If now we turn our attention to the oviparous spleen, we
find it to be actively engaged in the production of the self-same
bodies.   In ordinary preparations of the splenic pulp we see
only a mass of nuclei (such as are depicted in Photograph 179,
Plate XXII., which is a specimen obtained from the spleen of
the pike) more or less mixed up with red corpuscles derived from

the blood ; but when care is taken to get the specimens very thin, and in every way to preserve them as much as possible from *post-mortem* changes. we find that these nuclei belong to clear colourless ellipsoids such as are seen in Photographs 180 and 181, Plate XXII., which are also the products of the spleen of the pike. The former specimen was stained with a weak solution of aniline brown, and the latter by a stronger solution. It is only on the edge of the specimen in the latter that the corpuscles are seen plainly, but the mass is entirely formed of corpuscles of this character. In Photograph 182, Plate XXII., the same corpuscles are seen from the spleen of the perch. This specimen was stained with saline aniline blue, and, as a consequence, the nuclei are a little swollen. There is often a tendency in the clear margins to fuse together, and this allows the nuclei to come into contact, and then they also undergo fusion. These actions may be seen in progress in Photograph 183, Plate XXII. These colourless nucleated ellipsoids are developed from the comparatively opaque colourless corpuscles depicted in Photograph 184, Plate XXII.* I shall have occasion again to refer to these bodies in treating of the *minor* mode of the development of red nucleated corpuscles which goes on in the blood itself. In the above remarks I have confined myself exclusively to the *major* process of blood production in the ovipara as it is carried on in the spleen, bone-marrow, and lymphatics.

On the Minor Mode of Development of the Red Nucleated Corpuscle of the Ovipara.

I have elsewhere endeavoured to show that, in addition to the great blood-making process which exists in mammals, and which consists in the transformation of the most developed products of the various corpuscular sources into *colourless blood*

---

* These bodies are the analogue of the *primary* lymphatic and splenic corpuscles, and of the *nucleolated nucleus* of the bone-marrow of the mammal.

discs, there exists an arrangement by which the products which have failed to undergo full development in these sources may still be utilised and ultimately reach the goal for which they were originally intended.   I had long ago ascertained that this was true of the ovipara, at least in the case of the frog.

A corpuscle of the *primary order* * passes over from the spleen or bone-marrow and developes in the blood into a multi-nuclear white corpuscle.   In the ovipara these nuclei contain nucleoli, but the latter so fill them up as to be unrecognisable as distinct structures.   These bodies grow, and the capsule or membrane of the nucleus becomes distended so as to reveal the nucleolus within, and to form a cellular margin.   The nucleolus has by this time become oval, and this causes the body to take on the elliptical shape.   These bodies gradually obtain colour and become the red nucleated corpuscles.   In their slightly coloured stage they are the so-called hæmatoblasts of Hayem.   This minor mode of blood development is illustrated in Plate XXIII.   Although these bodies result ultimately in the formation of ordinary red nucleated corpuscles, the mode of development is very different to that by which the major proportion of the red nucleated corpuscles are produced, in their colourless stage, in the spleen and the bone-marrow.   This will be seen at once by reference to Photographs 194 and 195, Plate XXIII., and their descriptions, which show the colourless nucleated ellipsoids which have been, so to speak, thrown *ready-made* into the blood from the spleen and bone-marrow.†   Photograph 185, Plate XXIII., shows the

---

* *Vide* Section V., "On the morphological elements of the blood glands.

† When we examine the fresh blood of the frog or Triton, we invariably see numbers of oval bodies which look like free nuclei, and have much the same appearance as those seen in the red corpuscles. These bodies are in reality the nuclei of colourless ellipsoids, in many cases as large as the red corpuscles ; but their cellular margins are so colourless, smooth, and transparent, and their refractive index so akin to the liquid in which they lie, as to be wholly invisible, and incapable also of being photographed.   Photograph 193, Plate XXIII., illustrates this fact. Every one of the apparently naked nuclei here seen is surrounded by a clear hyaline margin having about the same area as that of the visible

primary corpuscles which gradually undergo development *in the blood itself* through successive stages, as seen in Photographs 186, 187, and 188, Plate XXIII., in which the division and formation of the nucleolated nuclei can be easily traced. The nuclei ultimately become set free and go through the changes previously referred to, which are depicted in Photographs 189, 190, 191, and 192, Plate XXIII. This mode of development has been very carefully investigated by M. Pouchet in the blood of the Triton ; but he has, I think, erred in considering the colourless stage of the nucleated ellipsoid and the white blood corpuscle to both arise out of the same body, and this a " naked nucleus." This origin from a nucleus appears to me to be true only of the colourless stage of the red nucleated cell. Many of the bodies which appear to be *naked nuclei* can be made by osmosis to show a cellular envelope. The white corpuscles appear to be developed from a primary cell, and the colourless stage of the red nucleated corpuscles from nucleolated nuclei, which are the offspring of these white corpuscles ; so that this primary cell, passing through the white corpuscle stage, runs its course, and has expended its powers when the *red nucleated corpuscle* has reached the period of its most perfect development, and its existence terminates in the degeneration of this latter body.

The colourless stage of the red nucleated corpuscle is preceded by a nucleus, or rather by a nucleolated nucleus, which has been set free from the white corpuscle ; but the latter body itself does not proceed from a nucleus, but from a primary splenic or bone-marrow cell, which, by being brought into the

---

red corpuscles. This margin may be rendered obvious by staining with fuschine (*vide* Photographs 194 and 195, Plate XXIII.), or by introducing some very fine power, such as Indian ink, into the blood, which attaches itself to these cellular margins and thus maps them out. I have found these corpuscles to be present in all classes of the ovipara, birds, fish, batrachia, and other reptiles. They are the ultimate or most developed products of their spleens and bone-marrows, and come over into the blood in a *colourless state*, there to acquire hæmoglobin.

blood before undergoing the regressive change necessary to the liberation of its nucleus (nucleolated), is stimulated by excess of pabulum to renewed cell growth and development, presenting itself in the first place as an uninuclear, and subsequently as a multinuclear white corpuscle. The stages in this multiplication of nuclei may be traced in Photographs 186, 187, and 188. Plate XXIII.

The development of the white and red corpuscle, as it occurs in the blood, proceeds, therefore, in one unbroken line, beginning with a primary lymph corpuscle, and ending with a fully-developed red nucleated corpuscle. The bodies recognised as the uninuclear and multinuclear white corpuscles are simply elements or links in this chain of development. This idea differs from that of M. Pouchet, who considers that the *same body* may take one or the other of two departures, one of which leads to the production of the white corpuscle, and the other to that of the red nucleated cell.*

---

* Les conclusions générales suivantes nous paraissent ressortir du travail qui précède :—

1. " Les hématies et les leucocytes chez les ovipares dérivent d'un seul et même élément anatomique.

2. " Le noyau des leucocytes subit une segmentation complète l'amenant à l'état d'amas nucléaire. Celui-ci est toujours concentrique à l'élément.

3. " La segmentation des leucocytes n'a jamais lieu tant qu'ils sont en suspension et en mouvement dans le sérum.

4. " Les prétendus faits de segmentation observés sur les leucocytes adultes en dehors des vaisseaux, ne sont que le partage (se produisant sous l'influence des mouvements du corps cellulaire) d'un amas de noyaux préalablement individualisés.

5. " Les hématies sont des formes élémentaires ultimes.

6. " Dans les hématies du Triton, le prétendu " réticulum " n'est qu'une apparence résultant du sectionnement partiel de la substance nucléaire.

7. " Le noyau de l'hématie atteint au cours de son développement un volume maximum, puis diminue jusqu' à la période d'état de l'élément.

8. " Les hématies disparaissent par dissolution dans le sérum ambiant.

9. " Il n'y a jamais chez le Triton de multiplication des hématies par scissiparie, dès que le corps de celles-ci a commencé de renfermer de l'hémoglobine.

What has here been said relates only to the *minor* process of blood formation. As before stated, we must carefully distinguish between the clear, colourless, nucleated ellipsoids, which are derived in a more or less fully developed state, so far as size and form are concerned, from the spleen and the bone-marrow, and those which are gradually formed in the blood, from nucleolated nuclei set free by the disruption of the fully-developed white corpuscle.

It is with this latter mode that M. Pouchet has concerned himself in the triton, and which appears to follow precisely the same law of development as in the frog.

The only point of difference between M. Pouchet and myself is that, whilst I consider with him that both the red and white corpuscles arise from *one* and the *same* body,* I hold their development to take place in an unbroken line, instead of by two diverse departures. M. Pouchet has fallen into this error by failing to recognise the *distinction* between the primary lymph and splenic corpuscles and the *nucleolated nuclei* † set free from the ordinary white corpuscles (bodies of the same size and appearance as each other), and it has landed him in the difficulty of having to consider that bodies of the *same kind which have the same origin as one another,* pass on to diverse issues. Inasmuch as they are all under the same external conditions, this is *primâ facie* a very unlikely thing, for such a divarication must either involve a difference in essential constitution, or in

---

" Enfin il existe peut-être une relation entre l'état moléculaire de l'hémoglobine existant dans les hématies (mais non telle que nous l'extrayons) et les deux formes régulières ovoïdes ou discoïdes, sous lesquelles celles-ci se présentent suivant les espèces animales."—M. Pouchet Journ. de L'Anatomie et de la Physiol., 1879.

* These bodies may, in fresh blood, always be distinguished from each other by the use of weak saline aniline blue, which stains all the nuclear bodies, leaving the primary lymph and their derivatives, the white corpuscles, for a long time unaffected.

† This body which M. Pouchet designates " Noyau d'origine," I regard as the *primary* lymph and splenic corpuscle and the corpuscle of the bone-marrow, which is equivalent to these. It is not a naked nucleus, but a cell with its capsule closely applied to the nucleus.

environment. As before pointed out, it is only with that mode
of development of oviparous blood which takes place in the
blood itself from a *colourless or white corpuscle* that we are here
concerned, and which is illustrated on Plate XX., Diagram II.,
under the head of the minor mode of development of oviparous
blood. The major mode of development of oviparous blood
from *colourless ellipsoids* formed in the lymphatics, spleen, and
bone-marrow does not appear to have been hitherto recognised,
certainly not as a distinct process of blood formation. This
mode is also illustrated on Plate XX., Diagram II., which may
be studied with advantage in connection with Table E.

Table E.

TABULAR VIEW OF THE MODES OF DEVELOPMENT OF MAMMALIAN AND OVIPAROUS BLOOD.

ANALOGUES AND EQUIVALENTS.

MAMMAL EMBRYO.

LYMPHATIC VESSELS AND GLANDS.	SPLEEN.	BONE-MARROW.	BLOOD.	
Primary Lymph Corpuscle	Primary Splenic Corpuscle	Freed nucleolated nucleus from pigment cell.	Colourless, clear-bordered, nucleated cell or developed white Corpuscle.	Partially coloured and full red nucleated cells. } Red disc.
Advanced Lymph Disc ...	Advanced Splenic Disc ...	Free nucleolus of marrow cell,	Colourless disc	Partially coloured and full red discs.

YOUNG AND ADULT MAMMAL.

LYMPHATIC VESSELS AND GLANDS.	SPLEEN.	BONE-MARROW.	BLOOD.	
Primary Lymph Corpuscle	Primary Splenic Corpuscle	Nucleolated nucleus from pigment cell.	Uni and multinuclear white Corpuscles.	Free nuclei or colourless discs, } Partially coloured and full red discs.
Advanced Lymph Disc ...	Advanced Splenic Disc ...	Nucleolus of smooth marrow cell.	Colourless disc	Partially coloured and full red discs.

YOUNG AND ADULT OVIPARA.

LYMPHATIC VESSELS AND GLANDS.	SPLEEN.	BONE-MARROW.	BLOOD.	
Primary Lymph Corpuscle	Primary Splenic Corpuscle	Opaque nucleolated nucleus from pigment cell.	Uni and multinuclear white Corpuscles.	Freed nuclei filled by nucleoli; the nuclear wall becomes distended, and the nucleolus smaller and more compact, and a colourless ellipsoid is gradually formed. } Partially coloured and full red nucleated ellipsoids.
Nucleated, clear-margined, colourless ellipsoid.	Nucleated, clear-margined, colourless ellipsoid.	Nucleated, clear-bordered, colourless ellipsoid.	Transparent, colourless, nucleated ellipsoid.	Partially coloured and full red nucleated ellipsoids.

On the Analogy and Relation which subsist between the Development of Mammalian and Oviparous Blood.

In addition to the great process of blood-making which goes on in the mammal, by which colourless discs derived from the glands, spleen, and bone-marrow are converted into red discs in the blood, I have ventured to suggest the existence of a numerically less important mode which is, as it were, secondary and supplementary to the first, and which utilised such elements as may escape from the blood-glands before they have undergone the peculiar changes which it is the function of these organs to bring about. I allude to the development of the *primary* lymph or splenic or bone-marrow corpuscle in the blood from a uninuclear to a multinuclear cell, and the probable shedding of these nuclei to become in their turn colourless blood discs. Certain facts which I have ascertained in the ovipara appear to be highly corroborative of this view, for we find that here also two distinct modes of development of the red nucleated corpuscle exist. A *major* one, which is that carried on by the spleen and the bone-marrow, and explained in a previous section, and a *minor* one, in which the development and formation of the corpuscle takes place in the blood itself. This latter mode I have illustrated by examples taken from the blood of the frog. It corresponds in all essential particulars with the method of formation of colourless discs in the blood from the nuclei of white corpuscles, with the exception that in the ovipara the bodies liberated are not simply nuclei, but nuclei containing large nucleoli. These nucleolated nuclei gradually develop, and, passing into the elliptical stage (Hayem's corpuscle), become red nucleated cells.

As an additional aid to the reader, I append Table E, which gives in a succinct form my views of the mode of development of mammalian and oviparous blood, and which may be studied with advantage in connection with Plate XX.

If read from left to right the bodies from each source, which are the natural *equivalents* of each other, are seen, and also the final product which they yield to the blood, and which, by becoming coloured, developes into the various kinds of red corpuscles. From above downwards, on corresponding lines, the *analogues* in the several kinds of blood are seen.

M. Hayem has sought to maintain the existence of perfect independence between the white and red corpuscles both in mammals and ovipara,[*] and, indeed, this is essential if the theory

[*] Mais c'est évidemment avec les globules blancs de la variété 1 que les hématoblastes peuvent être le plus facilement confondus. Nous résumerons dans le tableau suivant les caractères qui peuvent servir à différencier ces deux éléments.

GLOBULES BLANCS.	HÉMATOBLASTES.
Forme sphérique.	Forme plus aplatie et plus allongée.
Corps protoplasmique finement granuleux.	Corpuscule homogène.
Noyau homogène ou à peine nuageux, remplissant presque complètement l'élément, se colorant d'une manière très intense par la rosaline.	Noyau relativement moins volumineux, contenant des granulations disposées d'une manière particulière ; se colorant par la rosaline, moins fortement que celui des globules blancs.
Corps protoplasmique dépourvu, en général, de granulations brillantes.	Présence fréquente et constante chez certains animaux *(rana temporaria,* par ex.) de granulations brillantes probablement vitellines.
Élément se réduisant à l'état sec à une très mince pellicule tout à fait incolore, contenant un gros noyau homogène.	Corpuscule conservant à l'état sec sa forme et son volume, prenant un aspect vitreux et une teinte jaunâtre manifeste ; noyau presque indistinct.
Éléments isolés dans les préparations.	Éléments ayant la plus grande tendance à se grouper pour former des amas plus ou moins considérables.

Nous ajouterons comme dernière remarque que le globule blanc de la variété 1 n'est pas un élément particulier au sang de la grenouille et des autres ovipares. Il existe chez tous les vertébrés et possède, chez les animaux supérieurs, les mêmes caractères que chez les ovipares les plus infimes ; sa présence n'est donc pas liée à l'existence des hématoblastes à noyaux, et il n'a sans doute pas de rapports plus étroits avec les hématoblastes chez les animaux à globules nucléés que chez les animaux supérieurs.

of *hæmatoblasts* which he has propounded is to be sustained, for such a view renders it necessary to regard the lymph, splenic, and bone-marrow corpuscles as distinct elements which have nothing to do with the production of the red blood corpuscle.

I have, on the contrary, throughout this work, endeavoured to prove that correct views of the development of the blood are capable of utilising and harmonising the great body of physiological and pathological facts which past industry has accumulated by showing that the *lymphoid elements* from all sources must be regarded as either the direct or indirect progenitors of the *red corpuscles*.

GENERAL CONCLUSIONS.

1

That the true function of the blood-glands, or more properly speaking, of the lymphoid organs of the mammal (lymphatics, spleen, thymus, thyroid, bone-marrow, etc.) is to produce *colourless discs*, slightly less in diameter, and a little thicker than the blood discs. These discs are of two kinds, or exist in two stages—the cellular and the nuclear stage. The former I have designated the *primary*, and the latter the *advanced lymph disc*. The former is larger, more granular, and visible; the latter more translucent and less visible. Both these bodies pass over into the blood, but the latter in by far the greater numbers, being the true and final product of the lymphoid organs.

2

Passing to the blood side, we recognise again the presence of these lymph discs in small numbers, but they are now mainly of the primary or cellular variety, a few only of the free translucent nuclear discs being visible; but we are able to demonstrate that this is owing to the greater number of them having now become the invisible discs of the blood. These, subsequently, by the gradual acquisition of colour, again become visible, and present themselves as the pale intermediate forms of the blood discs, which ultimately become the full red discs. This is the major mode of development in mammalian blood. On the other hand, between the primary or cellular form of the lymph disc seen in the blood, every transition form up to the multi-nuclear white corpuscle can be found. The origin of both the red and the white corpuscle in the blood is therefore accounted for, and their relation to each other demonstrated.

3

The conversion of the primary lymph disc into the white blood corpuscle is simply an accessory and more protracted mode of development, apparently for the object of preventing formative waste, by utilising those products which come over from the lymphoid sources in an imperfectly developed state, for the nuclei of the white corpuscles being set free also pass through the invisible stage, and gaining colour, appear as pale blood discs. This method is described as the minor mode of blood making in mammals.

4

The function of the lymphatics, spleen, and bone-marrow of the ovipara or of their lymphoid organs is to produce colourless corpuscles, which also exist in two stages, a granulous lymphoid stage, and the more developed stage of a clear, colourless, nucleated ellipsoid, the margins or cell bodies of which are invisible, and which is the analogue of the advanced lymph discs of the mammal. To these bodies also the terms primary and advanced lymph corpuscles are applicable. As in the mammal, both these bodies pass over into the blood, but the latter in by far the greater numbers.

5

When we examine oviparous blood we find the primary corpuscles and a large number of bodies which appear to be free nuclei, but which on careful inspection, and by the use of various methods, are found to be ellipsoidal corpuscles (similar to those seen in the lymphoid organs), the margins of which are colourless and invisible. These are the analogue of the invisible discs of the mammal. Their marginal portions or cell bodies gradually acquire colour and become the pale red nucleated corpuscles. This is the major mode of development in oviparous blood.

6

On the other hand, the primary or granulous lymphoid corpuscles undergo development into the multinuclear white corpuscle, and its nuclei (which are the analogue of the nuclei of the white corpuscles of the mammal) being set free, are also seen to undergo gradual development and growth into colourless ellipsoidal cells, which attain colour before they have arrived at the size of the former variety. These are the bodies which have been described by Hayem as the hæmatoblasts of oviparous blood. They represent the accessory or minor mode of development of the oviparous red corpuscle.

7

As a lymphoid organ, the bone-marrow possesses special characteristics, which distinguish it from the spleen, lymphatics, thymus, etc., in the fact that, at an early period of embryo life, its colourless nucleated elements develop *in situ* into red nucleated cells, and that, at a more advanced period, these undergo conversion into, or give rise to, red discs; but even in the embryo, side-by-side with this mode, a new method for the formation of red discs is set up, which consists in the *colouration of naked nuclei*, which have been set free from the colourless cells of the marrow. This differs only from the general mode of blood formation I have described in the fact that colour is obtained in some instances before the nuclei leave the marrow, instead of after they have entered the blood. These embryonic modes of blood formation persist in some of the lower mammals throughout life, alongside the numerically far more important mode, in which free nuclear

discs leave the marrow and enter the blood in a *colourless state.*
In the higher mammals the bone-marrow at birth ceases almost
entirely to produce discs by means of red nucleated cells, and
in common with the other lymphoid organs, transmits to the
blood naked nuclei, or advanced lymph discs, which, passing
through the invisible stage, reappear as pale, red discs, and
ultimately become fully coloured. It is, of course, possible that
under some conditions embryonic modes may be reverted to,
and in such cases even the bone-marrow of higher mammals
might be found to again produce red discs from red nucleated
cells. These are not, however, its normal conditions. We
have already shown that the colourless ellipsoids of the ovipara
are the analogue of the advanced lymph discs of the mammal,
and, like these, they are transmitted to the blood in a *colourless
state* to obtain their hæmoglobin in the circulation.

8

As both the mammal discs and the oviparous ellipsoids
enter the blood in a *colourless state* and acquire colour while
circulating in it, the inquiry naturally arises as to the
source of their colouring matter. It is desirable to bear in
mind, as a possible solution of this problem, the fact that all
the lymphoid elements, without exception, while in their *primary
condition,* contain pigmentary matter, which is yielded to the
plasma of their several organs as development proceeds, and
which plasma, of course, passes with them into the blood. The
bone-marrow is particularly rich in this pigmentary matter, and
it is precisely here that cells and nuclei often become coloured
in situ, a fact which we might reasonably attribute to the
presence of a plentiful supply of pigment for conversion.

9

Leukhæmia is held to be due to a hyperplasia of the
lymphoid organs, inconsistent with that lengthened stay of the
products in these organs which is necessary to the decapsulation
of the primary lymph disc and the setting free of its nucleus as

the advanced lymph disc. The consequence of this is that the primary discs, which have a cellular constitution, pass into the larger lymphatics and blood vessels, and pursue their development as white corpuscles, instead of passing through the nuclear and invisible stage to form red discs by the direct or major process; leukhæmia is, in fact, the substitution of the minor or accessory mode of blood-making for the direct major mode, hence we can readily understand how in this condition increased production of white corpuscles is attended with diminution of red ones.

10

Anæmia, dependent upon disease, is regarded as being due to a diminution of the vitality of the lymph discs, which leads to a reduction of their life period, and therefore to decrease of their numbers; and inasmuch as the hæmoglobin is gained gradually in the blood, and in equal quantities in equal times, so the disproportionate decrease of hæmoglobin to corpuscles in the ratio of 3 to 2, as shown by experiment, is explained. Hence, given in any ordinary case of anæmia, the number of corpuscles per cubic millimetre, we can ascertain the average life-period of the discs and the average number of hæmoglobin increments per corpuscle for this life-period. This explains the *typical* variety. On the other hand, if the vitality of the corpuscle be differently affected, its hæmoglobin-producing function may become seriously impaired, and the hæmoglobin average be reduced below that proper to the number of corpuscles in the typical kind, and again, if the vitality is more profoundly influenced, then the number of discs formed by the lymphoid organs may be diminished, and the condition induced known as pernicious or essential anæmia.

11

It has been sought to show that fibrin formation and coagulation is not so entirely a chemical problem as hitherto supposed, but that so far as the lymph and blood are concerned, the process rests upon a purely morphological basis, in the former case this action being entirely ascribable to changes

which occur in the advanced lymph discs, and in the latter to
the younger blood discs which are in fact simply the former
bodies in a more advanced stage. Coagulation as seen in the
lymph and blood therefore depends upon those discs, which, on
account of their proneness to change when the blood is shed,
have been designated in this work the *fugitive group of discs.*
This comprises the nuclear discs which commence with the
advanced lymph corpuscle, and end with the green, lustrous,
diffused edged blood disc. The bodies which Hayem has named
hæmatoblasts in mammal blood, and which he has shown to be
associated with fibrin formation, possess these properties,
simply by virtue of the fact that they are modified
forms, and fragmentary portions of the younger blood discs,
of those in fact, which have already acquired a little
colour. The blood corpuscles of the ovipara are, equally
with those of the mammal, susceptible of division into a
permanent and a fugitive group, the former being represented
by the more fully coloured corpuscles, and the latter by the
colourless and slightly-coloured ellipsoids derived from the
lymphoid organs, and the young ellipsoids (so-called hæmatoblasts)
which develop and grow in the blood itself, the clear pellucid
margins or cell bodies being the portions which in various ways
undergo modification and conversion into fibrin. This view
does not interfere with the chemical views of fibrin-formation so
far as pure solutions containing the separate elements necessary
to coagulation are concerned; but, on the contrary, throws great
light on the nature of these processes, and on the relation
which the substances, known as fibrinogen, fibrinoplastin, and
ferment bear to the constituents of the corpuscles.

MICRO-PHOTOGRAPHS OF SPECIMENS

AND THEIR DESCRIPTIONS.

DESCRIPTION OF PLATES.

PLATE I.*

Fig. 1 shows the arrangement for producing, by means of Newton's rings, a barrier which prevents the passage of the blood corpuscles, while allowing that of the liquor sanguinis. The blood being inserted at A, the serum filters off into the space B. As a consequence, the corpuscles become packed together in the space A, and the *colourless discs* are displayed among them.

Fig. 2.—An arrangement of the slide and cover glass, by means of which the earliest matters that adhere can be readily examined in the absence of the obscuring influence of the liquor sanguinis, which can be removed and restored at will by gentle movement of pinion screw. (*Vide* page 5.)

Fig. 3.—Arrangement for getting rid of the surplus blood by capillarity, it admits of the adhering corpuscles, granules, &c., being exposed to the influence of fixing vapours or solutions (osmic acid, &c.) the instant they are removed from the liquor sanguinis. (*Vide* pages 8 and 10.)

Fig. 4.—Arrangement for removing surplus blood by means of air or other gases, charged with dried vapours. (*Vide* page 6.)

* All the corpuscles up to Plate XI. are magnified 476 and the remainder 500 diameters. A micrometer scale divided into 10 and 20,000ths of an inch is given with each magnitude. (*Vide* Plates IX. and XXIII.)

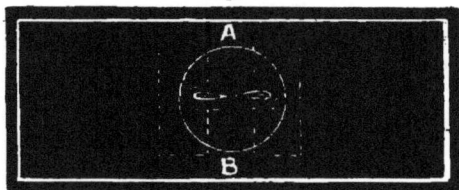

Fig. 1.

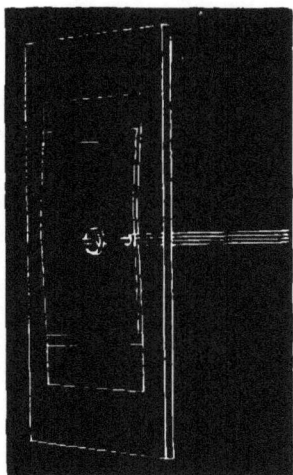

Fig. 2.

Fig. 3.

Fig. 4.

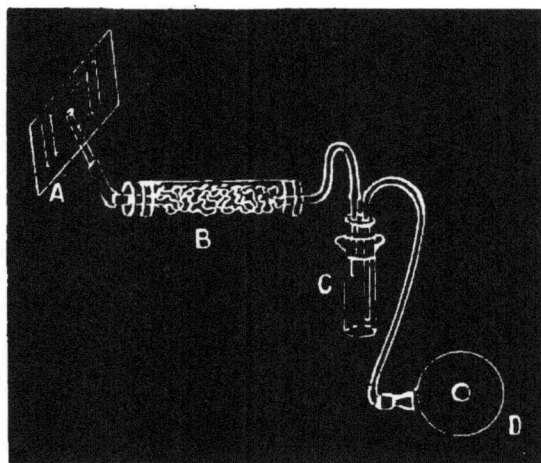

PHOTO. 1.

PHOTO. 2.

PHOTO. 3.

PHOTO. 4.

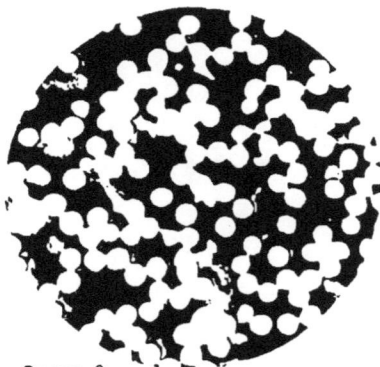

PHOTO. 5.

PHOTO. 6.

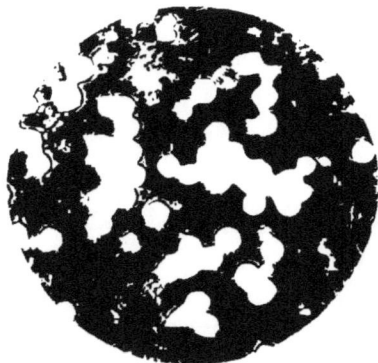

PLATE II.

Photograph 1.—Ordinary red biconcave discs, the form of which has been preserved by osmic acid vapour.

Photograph 2.—Ordinary white blood corpuscles. The red discs present have in this preparation become spherical, and look, consequently, smaller and darker.

Photograph 3. — Specimen of blood from which the liquor sanguinis has been *partially* withdrawn. The dark masses consist, in some cases, of single, deformed red corpuscles, and in others of several fused together. Many of these masses will be seen to be bounded by curvilinear indentations, and if some of these are carefully examined, it will be observed that they are due to the presence and influence of certain disc-like annular bodies, which are either pressing into the red corpuscular masses, or forming a basis around which these plastic masses are applying and adapting themselves. The latter view is probably the correct one, inasmuch as the red corpuscular masses may often be seen to swim freely about in the liquor sanguinis, still retaining their concave outlines, and, fitting into some of these, delicate, nearly invisible corpuscles may be detected. This being the case, it is fair to consider that the concavities in the cases where no corpuscles can be seen are due to the presence of corpuscles which are quite invisible.

Photograph 4.—In this specimen the liquor sanguinis has been filtered off to such an extent that the interstices between the invisible corpuscles have become filled with closely packed red corpuscles, and these, forming a new background, reveal the *apparent spaces* in which the invisible corpuscles lie. It will be seen that, in some instances, these corpuscles have attained a slight shade of colour, and, therefore, can be seen ; but in the majority of cases the light is as easily transmitted as through the glass itself. Some of the colourless discs are fusing together, and others are connected by a small neck of their own substance. These two facts show their liquidity.

Photograph 5.—In this specimen, the proximity of the *red* corpuscles is less complete, and the individual red corpuscles forming the dark background and revealing by contrast the invisible discs can be distinguished.

Photograph 6.—A preparation in which the outlines of the invisible corpuscles have been made just apparent where they are in contact with and press against each other by altering the refractive index with saturated solution of salt.

PLATE III.

Photograph 7 shows the influence of long-continued action of salt. The invisible corpuscles become contracted and their edges stained by hæmoglobin derived from the coloured discs, which are also condensed.

Photographs 8, 9, and 10 display groups of young corpuscles, which have been withdrawn from the blood by means of their adhesiveness to solids. As the degree of adhesiveness is inversely as the colour, it is not difficult to withdraw from the blood a large number of its younger corpuscles, for these readily attach themselves to the slide or the cover-glass; the best method of doing this is to fasten down the cover-glass at one end with plaster (forming a sort of hinge) upon a slide which has been drilled, and has a fine screw working in a bush, by which the cover-glass can be gently raised at the opposite end to which it is attached, and the mass of blood be made to recede by capillarity to the other end. This requires to be done with the utmost gentleness, for if the movement of the blood is too great the youngest corpuscles are broken up into granules. About four minutes should be allowed for the corpuscles to adhere.

Photograph 11.—In the centre of this group four of the colourless discs have become fused together into a smooth mass. At the broad end of the group there are two similar masses. That on the left has a young, slightly coloured corpuscle incorporated in it. Such masses as these frequently granulate.

Photograph 12.—In this preparation we see colourless discs isolated from the groups. Three of these are in the smooth state, but a fourth on the right is undergoing granular disintegration. Various granular masses and particles are also seen. These represent *débris* of the colourless discs. This mode of degeneration is one of the numerous proofs that these corpuscles are not decolourised red discs. Lower on the right invisible corpuscles may be seen lying amongst red corpuscles, and being indented by them so as to give the appearance of spaces filled with mere white liquid matter.

PHOTO. 7.

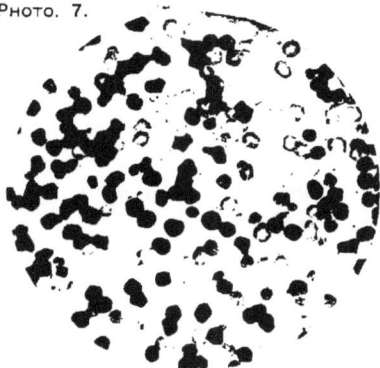

PHOTO. 8.

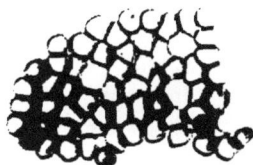

PHOTO. 9.

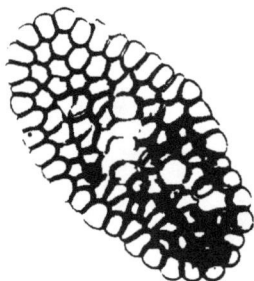

PHOTO 10.

PHOTO. 11.

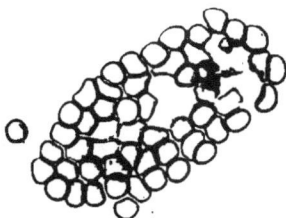

PHOTO. 12.

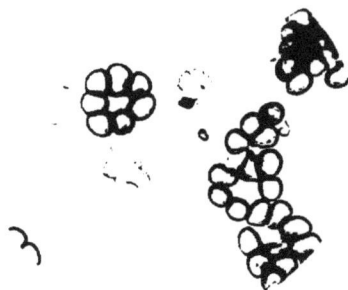

PLATE IV.   *Colourless and Intermediately Coloured Discs, etc.*

PHOTO. 13.

PHOTO. 14.

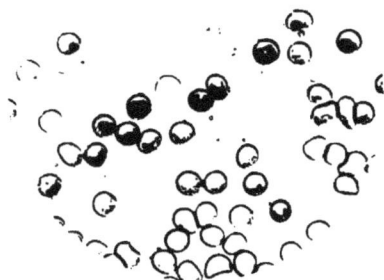
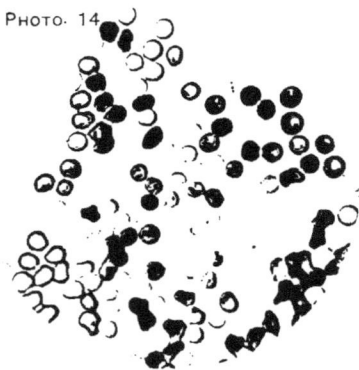

PHOTO. 15.

PHOTO. 16.

PHOTO. 17.

PHOTO. 18.

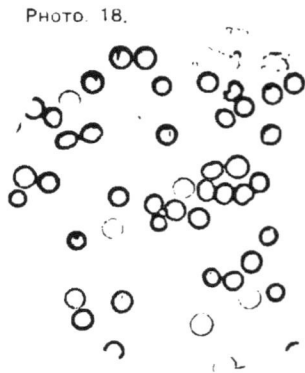

PLATE IV.

Photograph 13 shows numerous isolated colourless discs and some that have attained faint traces of colour.

Photograph 14.—In this preparation the invisible discs appear to be absent, while there are present many partially coloured ones—colourless discs are, however, in this case lying like liquid in the capillary spaces between the coloured discs. This is well seen in a group at the left-hand side of the specimen. Small fragments of the matter of which the coloured discs are formed may be seen uniting two or more coloured corpuscles together. When these discs become separated from each other such little beads are drawn out into fibres. This is one way in which fibrin is formed.

Photographs 15 and 16 are examples of preparations obtained by practising the 'method of isolation' over the vapour of osmic acid. By this means all the various grades of the blood discs are preserved and fixed. But here, as in the previous case, some of the more delicate colourless ones lie like liquid in the capillary spaces between the more advanced ones, and even those which have attained considerable colour are seen to have spread themselves out, and to have acquired a greater diameter than the more coloured ones. Photograph 16 exhibits several nearly colourless discs in which the biconcave form has been retained.

Photographs 17 and 18.—Preparations of blood reduced to the freezing temperature, and subsequently exposed at this degree of cold to the dry vapour of osmic acid. By this means the biconcave form of the invisible and the slightly coloured discs can be readily preserved.

The granules seen in these specimens result from the breaking up of some of the partially coloured discs. They are the bodies which Hayem has described as the hæmatoblasts of mammalian blood. When these discs undergo granulation they do not invariably break up at once into separate and independent granules, but assume the character of a coloured granular corpuscle. In this condition (*vide* Photograph 18), they have been described by Semmer under the name of granule spheres, (*Körnerkugeln*,) as intermediate or transition forms beween the ordinary white corpuscle and the red disc.

PLATE V.

Photograph 19 shows a mosaic group of coloured corpuscles, surrounded by invisible ones. A keen sight may detect in every part of the background of this specimen dark lines, they indicate the existence of a layer of phantom corpuscles, which may be brought out by protracted staining with alcoholic solution of aniline brown. (*Vide* Photograph 20.)

Photograph 20.—This specimen has undergone protracted staining with aniline brown, and a complete background of corpuscles, which were previously invisible, has thus been brought into view.

Photograph 21.—This is a similar specimen to 20, with the difference that many of the invisible corpuscles have become already fused into liquid masses, which appear to adhere to and surround the more coloured corpuscles. In many cases the invisible corpuscles are still sufficiently distinct to show how the " plasmine pools " originate. There is reason to consider that every part of the background of this specimen is covered with liquid of corpuscular origin.

Photograph 22 shows invisible and slightly coloured corpuscles undergoing fusion, spreading, and disintegration. This represents a still more advanced stage in fibrin formation.

Photographs 23 and 24.—These specimens are introduced to show how the colourless and slightly coloured discs may be preserved by spreading and *rapidly* drying the blood. In this way we fix them before they have sunk down as a homogeneous layer upon the slide. The more rapidly the blood can be dried, the greater will be the number of colourless discs preserved. 23 represents spontaneous, and 24 rapidly-dried blood. (*Vide* page 10.)

PHOTO. 19.

PHOTO. 20.

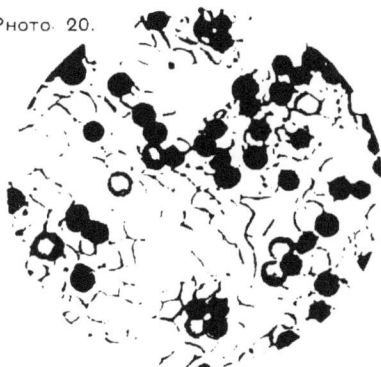

PHOTO. 21.

PHOTO 22.

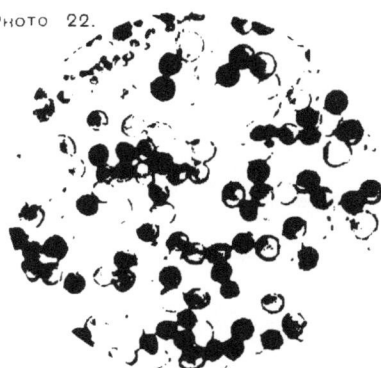

PHOTO. 23.

PHOTO 24.

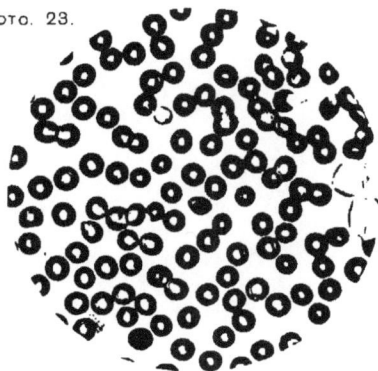

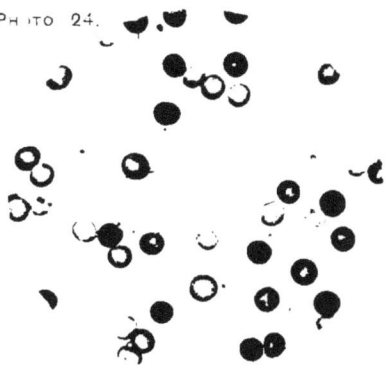

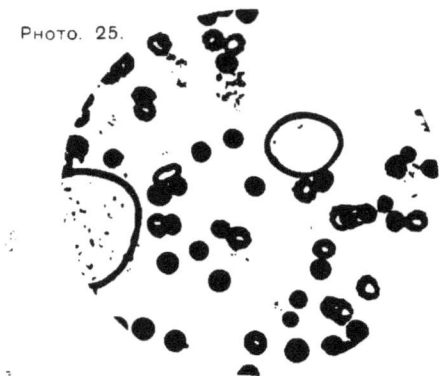

PHOTO. 25.

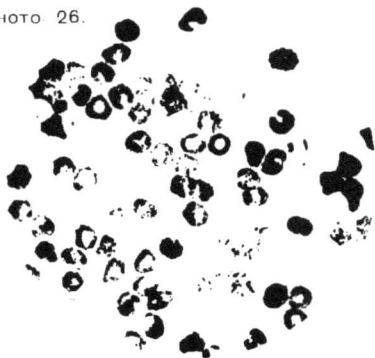

PHOTO. 26.

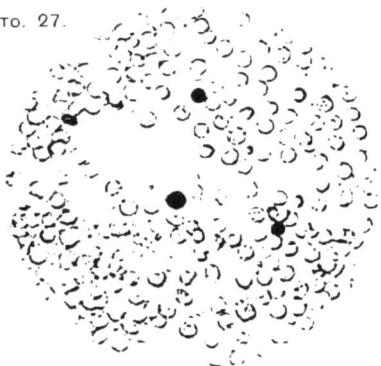

PHOTO. 27.

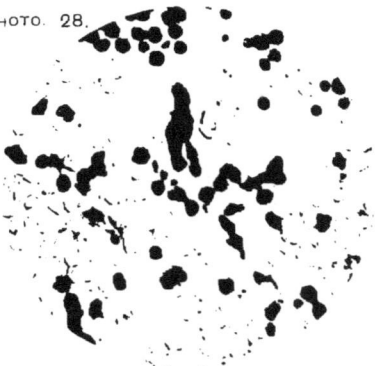

PHOTO. 28.

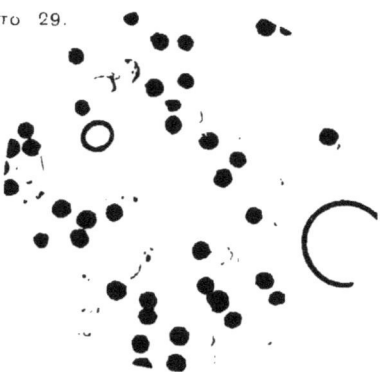

PHOTO. 29.

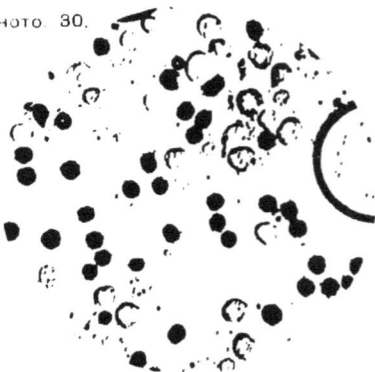

PHOTO. 30.

PLATE VI.

Photograph 25 is a specimen in which the invisible corpuscles have been brought into view in the serum by staining with carmine ; other methods have since been devised which are less difficult and in many respects preferable. (*Vide* Section III., page 66.)

Photograph 26.—Specimen showing dark, full red granules, frequently seen in fresh blood, and which appear to be the *débris* of red corpuscles which have reached the period of their full development, and are undergoing disintegration. This is probably one mode in which the red corpuscles finally disappear. They are also present in Photograph 23.

Photograph 27.—A specimen of lymph discs, taken from a lymphatic gland immediately after death. The dark, circular bodies here and there are red corpuscles. Some of these lymph discs may be seen to be already biconcave, and others are in process of change. The remarkable regularity of size of these bodies among themselves, and their easy gradation to the biconcave colourless disc, will arrest the attention of the observer.

Photograph 28 is a specimen of similar corpuscles from the spleen. They are extremely fragile, and liable to change, but when properly preserved are bodies closely akin in size and constitution to the gland corpuscles before described. The dark bodies and masses are red corpuscles.

Photographs 29 and 30 show the changes of size and appearance which frequently occur in the previous bodies, after the organs (glands or spleen) have been allowed to cool to the temperature of the room. The rings with dark outlines on these specimens are, of course, air bubbles.

PLATE VII.

Photograph 31 represents lymph corpuscles which have been submitted to the action of water, and subsequently stained on account of their extreme faintness. It will be noticed that they are very differently affected. Some of them have attained a diameter of 15-20,000ths of an inch, while others are scarcely swollen at all.

Photograph 32.—Changed appearance of gland corpuscles from aggregation and fusion. Their greater density and opacity distinguishes them from the merely swollen states of these bodies.

Photograph 33 shows more complete coalescence of the lymph corpuscular substance.

Photograph 34 is an example of fibrillation among lymph corpuscles. The corpuscles adhere to each other, and are pulled out into threads or fibres.

Photograph 35 shows a plate of lymph fibrin produced by the coalescence and melting down of lymph corpuscles into a continuous soft film upon the slide.

Photograph 36.—Lymph corpuscles, some of which may be seen to be surrounded by a light zone. (*Vide* Photographs 122, 123, and 124, Plate XVI.)

Photo. 31.

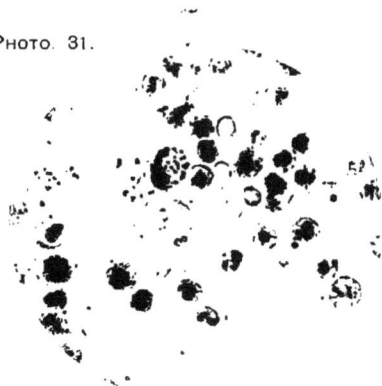

Photo. 32.

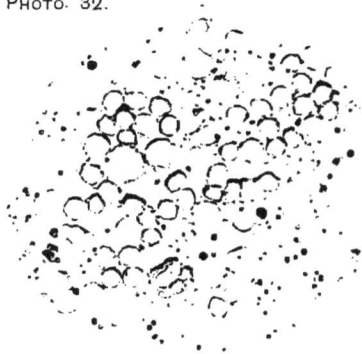

Photo. 33.

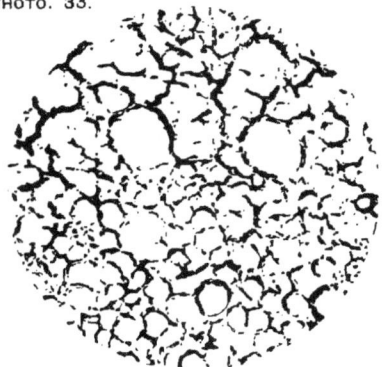

Photo. 34.

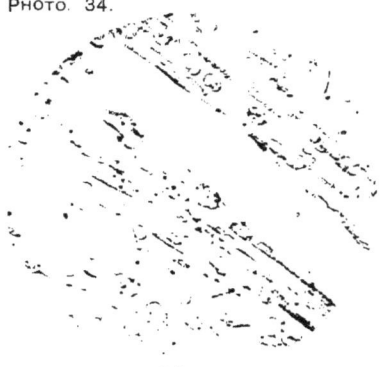

Photo. 35.

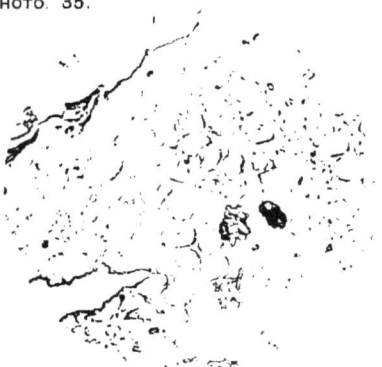

Photo. 36.

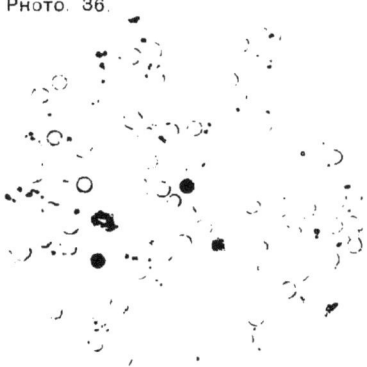

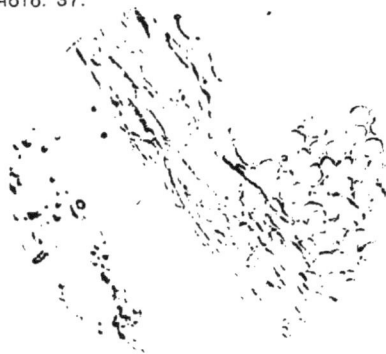
PHOTO. 37.

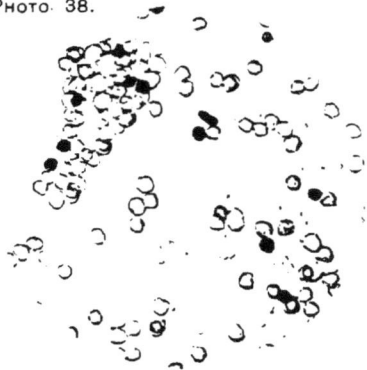
PHOTO. 38.

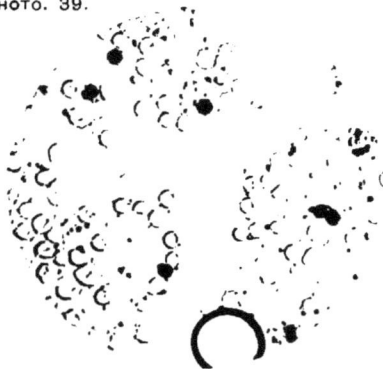
PHOTO. 39.

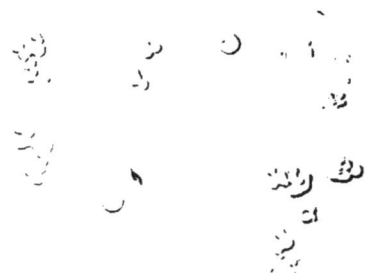
PHOTO. 40.

PHOTO. 41.

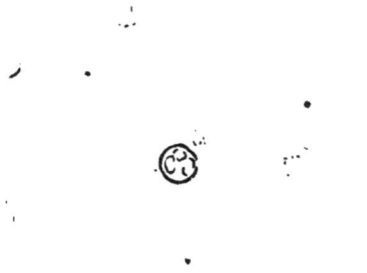
PHOTO. 42.

PLATE VIII.

Photograph 37.—Lymph corpuscles coalescing and giving rise to a thick liquid which flows slowly down the slide.

Photograph 38.—Lymph corpuscles which have been rendered irregularly spherical by two per cent. solution of salt. The dark bodies among them are spherical blood corpuscles.

Photograph 39.—Specimen of lymph discs, some of which have become smoothly bi-concave for about half or two-thirds of their extent, while the remaining portion still retain the usual regular aspect.

Photograph 40.—Biconcave discs of various sizes formed artificially from simple organic substances. (*Vide* Part IV., Section I., "On the causes of the bi-concave form of the mammal blood corpuscle.")

Photograph 41.—Myelin liberated from lymph corpuscles by means of strong acetic acid. The biconcave form of these little masses is worthy of notice.

Photograph 42.—A white corpuscle, the delicate capsular pellicle of which has been removed, exposing the nuclei and the smooth liquid protoplasm in which they are imbedded. (*Vide* Photograph 133, Plate XVII.)

PLATE IX.

Photograph 43 shows the discharge of the protoplasmic contents and the simultaneous contraction of the white corpuscle, caused by the action of a saturated solution of salt.

Photograph 44.—A specimen in which the nuclei also appear to have undergone rupture and contraction. The white content matter is here seen more plainly because it lies on a background of decolourised red corpuscles.

Photograph 45.—Granules resulting from the breaking up of younger or fugitive corpuscles.

Photograph 46.—Coagulum of pig's blood, on the surface of which may be seen granulating young coloured corpuscles, mistaken by Semmer for intermediate ones between the ordinary white and the red corpuscle.

Photograph 47 shows that some of the red blood corpuscles, if not all, have two constituents, a coloured and a colourless, the latter being masked or concealed within the former. They can be disengaged from each other in several ways. (*Vide* Photographs 67, 68, and 69, Plate XII.)

Photograph 48.—Micrometer divided into 10,000ths and 20,000ths of an inch. Objects in any of the photographs may be measured and compared with each other by means of a pair of compasses.

PHOTO. 43.

PHOTO. 44.

PHOTO. 45.

PHOTO. 46.

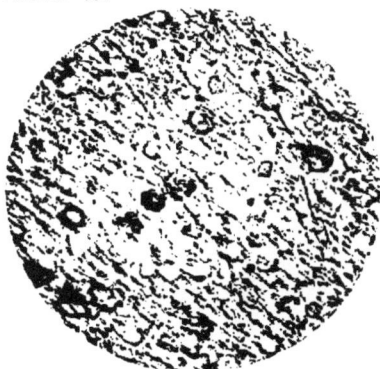

PHOTO. 47.

PHOTO. 48.

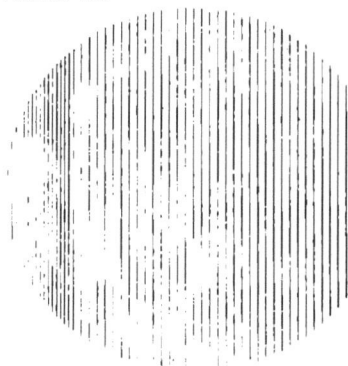

PLATE X. *Invisible Discs and Decolourised Red Discs.*

PHOTO. 49.

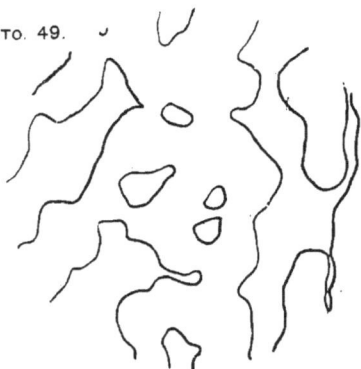

PHOTO. 50.

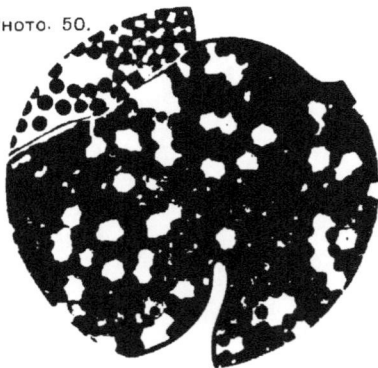

PHOTO. 51.

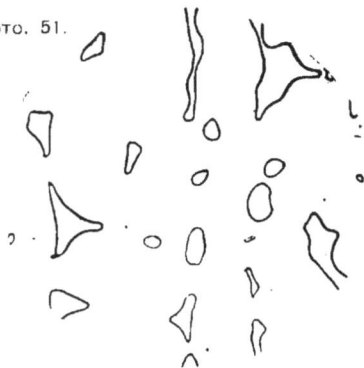

PHOTO. 52.

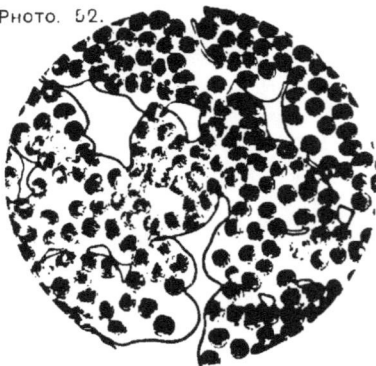

PHOTO. 53.

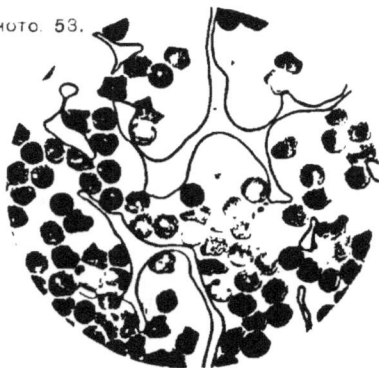

PHOTO. 54.

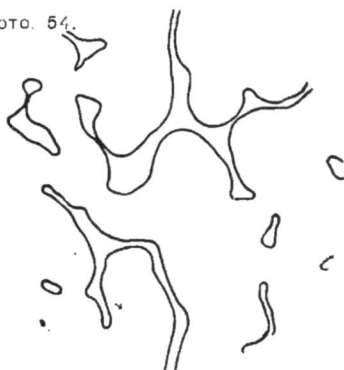

PLATE X.

Photograph 49.—Exhibits the differences of colour between the liquor sanguinis of blood freshly shed and of the spaces which contain air only, the difference of tint under these circumstances is usually very slight.

Photograph 50.—From a specimen in which the colouring matter has been prevented leaving the corpuscles by the use of a colloid (dry soluble albumen.) The air spaces, the liquor sanguinis, and the invisible corpuscles, are here all alike colourless.

Photograph 51.—Taken close to the rings of Newton. The liquor sanguinis is of a deeper tint than the air-spaces, owing to the presence in it of much hæmoglobin. The corpuscles are increased in diameter, and have a faint appearance, owing to the fact that they are made very thin by the proximity of the glasses, and because the contrast between them and the liquor sanguinis is diminished by the discolouration of the latter.

Photograph 52.—Another specimen near to the rings of Newton, in which red discs have been prevented from giving up their colour by the use of a colloid (albumen.) The serum is in this case nearly as colourless as the air-spaces. Photographed eighteen hours after preparation.

Photograph 53-54 represent precisely the same spot near to the barrier or rings of Newton. 53 shows the state of the specimen soon after preparation, and 54 after an interval of eighteen hours. No preservative was used. All the red corpuscles have disappeared *in situ* while the colourless discs which lay among them have remained, and may be seen as phantom forms in the coloured serum. If these were, as some have thought, decolourised red corpuscles, they should have been the first to disappear.

PLATE XI.

Photograph 55.—Groups of the invisible corpuscles in the spherical state obtained by the ' method of isolation.'

Photograph 56.—Similar groups to 55, fusing into white masses.

Photograph 57.—Shows the fusion of the colourless discs in a more complete state.

Photograph 58.—Here colourless discs have not only become fused but the resulting masses have commenced to granulate.

Photograph 59.—A specimen of the red disc, which occupies a mid-position between the corpuscles of the permanent and the fugitive groups. When seen under the microscope it has a lustrous flickering character, and is of a greenish colour. Like those of the " fugitive group," it has a disposition to fall down and spread itself like a liquid upon the slide.

Photograph 60.—An example of the *direct* conversion of the colourless discs into fibrin without passing through the stage of granulation. It is due to annulation modified by currents. (*Vide* Part IV., Section I.)

PHOTO 55.

PHOTO. 56.

PHOTO. 57.

PHOTO. 58.

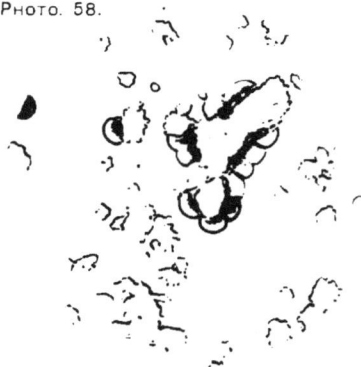

PHOTO 59.

PHOTO. 60.

61.
62.
63.

64.
65.
66.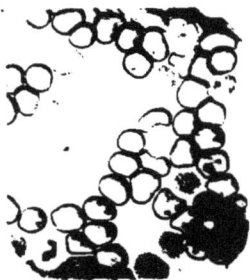

67.
68.
69.

70.
71.
72.

PLATE XII.*

Photograph 61.—A group of corpuscles withdrawn from fresh human blood by the "method of isolation" over osmic acid vapour. Examples are here found of every kind of blood disc, from the colourless to the full red stage.

Photograph 62.—We have here three colourless discs, well isolated from the red ones. A fourth to the right hand has become granular, and has lost its clearness of outline. On the left of the little group of red corpuscles there is a smooth disc, which is almost invisible, although the serum is absent. Masses which represent fused and broken-up discs in the granular state are also to be seen in the specimen.

Photograph 63.—Colourless discs in the early stages of granular disintegration.

Photograph 64.—Colourless discs in a more advanced state of fusion and granulation. The granules resulting from the disintegration of ordinary white corpuscles are finer and more distinct. (Compare Photograph 77, Plate XIII.)

Photograph 65.—Granules derived from the breaking up of colourless discs entering into the formation of fibrin.

Photograph 66.—In this specimen among the ordinary red discs may be seen others of the variety described in Photograph 59, Plate XI., as the green, lustrous-looking corpuscles, which occupy a position between the permanent and fugitive groups of the red corpuscles. These corpuscles are here seen to be undergoing change. White matter is appearing at their circumference, while the coloured material is collecting in their centres. These changes are the same as those previously described as taking place in the *colourless discs* (*vide* Photographs 62 and 63); but they are occurring now in corpuscles which have become coloured to a certain depth only, and not entirely through their substance; hence, when disintegration occurs, the material which has been converted into hæmoglobin is drawn away and condensed, leaving the colourless substance or part exposed to view. More extreme results

---

* All these specimens, from Plate XII. to XXIII., are magnified 500 diameters. The micrometer scale for this magnitude will be found on Plate XXIII.

PLATE XII.—*(continued).*

of this change, both in entire corpuscles and granules, are shown in Photographs 67, 68, 69, and 72.

Photograph 67.—Rosette corpuscles due to a more excessive action of the changes described in Photograph 66.

Photograph 68.—Rosette granules. The granules which arise from the breaking up of the young coloured corpuscles undergo precisely the same separation into coloured and colourless portions, as the entire corpuscles do. In the lower part of this specimen we notice a group of granules from which the intermediate white substance has nearly disappeared. Compare this with the groups of granules on Photograph 72. This white substance is the so-called cement matter referred to by M. Hayem. It is in reality the interior substance of young partly-coloured corpuscles and their granules.

Photograph 69.—Rosette corpuscles obtained from human blood which had been kept for a short time at its normal temperature by means of the warm stage. The white borders of specimens thus prepared are more spiked or stellate, and have a greater disposition to granulate.

Photograph 70.—Young partly-coloured corpuscles breaking up into granules, which are held together more or less by the interior white matter. Sometimes a corpuscle may be seen to break up *in situ*, and the granules may then be counted as in one example in this specimen. Corpuscles similar to those which have been broken up may be seen unaffected in the specimen.

Photograph 71.—Circlets and beaded rolls of granules described by M. Hayem as wreaths, garlands, and chaplets of so-called hæmatoblasts. Their origin is seen in Photograph 70.

Photograph 72.—Examples of granule masses (so-called groups of hæmatoblasts) held together by white substance of interior, which tends to fuse into a common mass (*vide* descriptions of previous photographs). The entire Plate illustrates the source of the so-called *hæmatoblasts*.

73.  74.  75.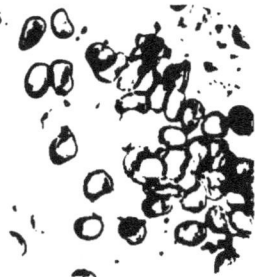

76.  77.  78.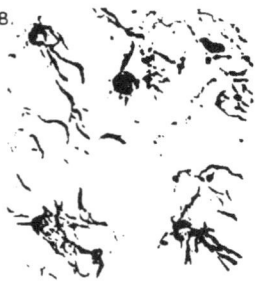

79.  80.  81.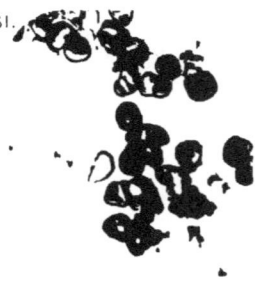

82. 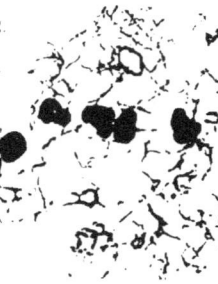 83.  84.

PLATE XIII.

Photograph 73.—In this specimen the coloured matter usually covering the exterior of the corpuscle, and hiding its white central part, has arranged itself as a thin zone around it, thus allowing it to come into view. The coloured part frequently becomes dissipated, and the white interior masses then coalesce forming groups. These fuse more closely, and homogenous white masses or sheets are formed, such as the one seen in the upper part of the specimen.

Photograph 74.—The large mass in the centre of this specimen is partly in the groupal and partly in the fused or homogenous state, which demonstrates that it is in this manner some of the smooth sheets and masses of white matter are formed.

Photograph 75.—In this specimen young corpuscles may be seen to be breaking up into granules, which become converted by extension into fine filaments of slightly coloured fibrin. This is the kind of fibrin which makes its appearance in an ordinary specimen of blood between the rolls. It can be seen in the serum, because it is slightly coloured.

, Photograph 76.—This specimen illustrates a curious effect which is constantly taking place in blood submitted to examination on glass slides. In the first place corpuscles and groups of corpuscles become attached to the surface, and over these threads of fibrin are formed, due to the extension of the viscous matter of the younger corpuscles by the red ones, which first effect slight adhesion to them, and subsequently, by swimming freely about in the serum, draw them out into fibres. When this network of fibrin contracts, as it shortly does, it cuts the corpuscles which are beneath it into fragments, and the result is a free production of so-called haematoblasts. (Traces of this action may be seen in Photographs 75 and 81.)

Photograph 77.—Granules resulting from the disintegration of three white corpuscles. These granules do not enter into the formation of fibrin.

Photograph 78.—This specimen shows entire corpuscles undergoing direct disintegration into fibres. In some cases the corpuscles are wholly resolved into a fibrous network ; in other cases the action is partial only.

PLATE XIII.—*(continued)*.

Photograph 79.—First steps in the formation of radiating fibres from the rosette corpuscles. The rosette corpuscles, masses, or granules, adhere to the surface of the slide, and red corpuscles frequently adhere to and around them. The white borders of the isolated ones readily spread down upon the slide and become lost to view, while coloured fibres are drawn out from the central dark parts by adhesion of the free corpuscles, as before indicated.

Photograph 80.—The action of fibrillation is in this case more advanced, the white borders of the rosettes have almost entirely disappeared, and the central dark portions are irregularly stellate and angular, owing to fibres being dragged out of them. These angular, central dark parts have been described by M. Hayem as hæmatoblasts. The white borders are invisible in the presence of the liquor sanguinis.

Photograph 81.—In this specimen the white borders have entirely disappeared, and the central dark masses are disintegrated into granules, and, becoming extended into fibres, many of these threads lie across red discs and are cutting them into pieces.

Photograph 82.—Some of the young corpuscles, without undergoing any preliminary disintegrative changes, have fibres drawn out from the substance of their periphery. In this way they gradually become lessened and exhausted. Such corpuscles have also been described by M. Hayem as hæmatoblasts.

Photograph 83.—A group of oviparous blood corpuscles, showing every stage of colour between a perfectly colourless nucleated and a full red nucleated corpuscle. There is in respect to colour a perfect analogy between the development of the mammal and oviparous corpuscle. (Compare Photograph 61, Plate XII.)

Photograph 84.—Colourless nucleated corpuscles, brought into view by staining. When such corpuscles are submerged in the serum their nuclei alone are seen, the margins being invisible, and therefore offering a perfect analogy to the invisible discs of the mammal.

PLATE XIV.   *Hæmatoblasts (so called), Blood Development, etc*

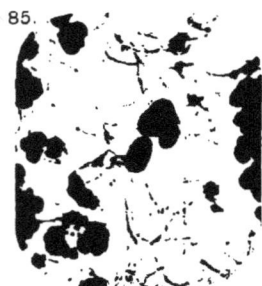

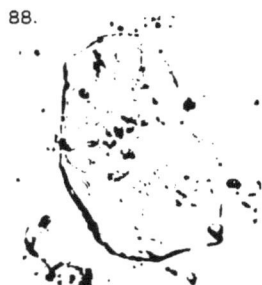

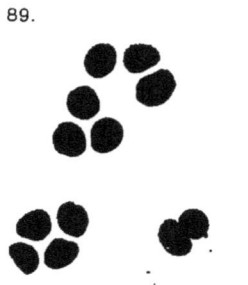

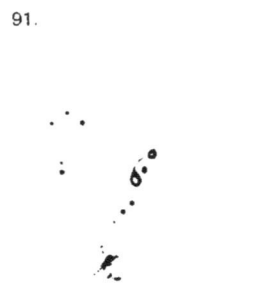

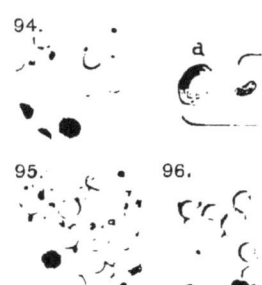

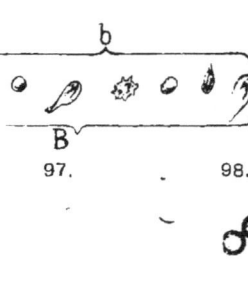

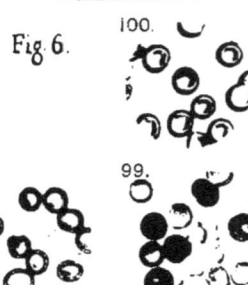

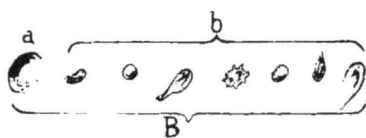

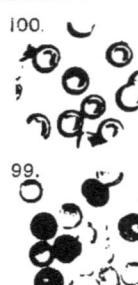

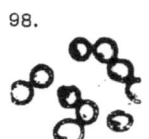

PLATE XIV.

Photograph 85.—An excellent illustration of the formation of fibrin from the periphery or exterior of young partially-coloured corpuscles, without disintegration of the corpuscles into their two substances, coloured and colourless, so-called hæmatoblasts.

Photograph 86.—Formation of radiating fibrin from the coloured substance of fragments of young discs (so-called hæmatoblasts.) The two materials of the corpuscles have in this case become separated from each other, and the white matter has gathered itself up into distinct globules, and does not seem to be sufficiently viscous to yield fibres, or is not adhesive to the *free* red corpuscles.

Photograph 87.—A mass of granules derived from the disintegration of comparatively colourless discs, two of which are seen on the margin of the mass.

Photograph 88.—A sheet of fibrin formed by coalescence and contraction from a similar mass of granules to those seen in Photograph 87.

Photograph 89.—Young highly-coloured corpuscles still retaining the tendencies to swell and spread possessed by the diffused-edged corpuscles. (*Vide* description of Photograph 66, Plate XII.)

Photograph 90.—Granular masses yielded by highly-coloured young corpuscles, so-called hæmatoblasts.

Photograph 91.—Another method in which coloured corpuscles divide up into the smaller bodies, which have been regarded as growing blood discs. The fragments are often biconcave in form. Microcytes may arise in this manner.

Photograph 92.—Young discs in the act of granular degeneration.

Photograph 93.—More highly coloured corpuscles undergoing granular disintegration.

Figure 6.—Diagram furnished by M. Hayem, illustrating his views of the development of the red discs from so-called hæmatoblasts.

PLATE XIV.—*(continued.)*

Photographs 94, 95, and 96.—Lymph corpuscles from the lymphatic glands of young pigs, just killed.   They are seen to be discs in various stages of development, and some of them have already taken on the biconcave form.   This is particularly noticeable in photographs 95 and 96.

Photograph 97.—Colourless discs obtained from fresh human blood.   Their biconcave form has been preserved by the combined influence of cold and osmic acid vapour.

Photographs 98, 99, and 100.—Specimens of human blood preserved with osmic acid vapour.   Among them may be seen all intermediate stages of colour between the colourless and the full red disc.   The more coloured corpuscles are a little contracted by the strength of the osmic acid necessary to preserve the younger and less coloured discs.   For this reason the concavities in specimen 100 are a little exaggerated.

101.

102.

103.

104.

105.

106.
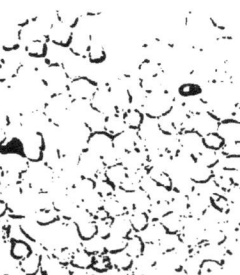

107.

108.
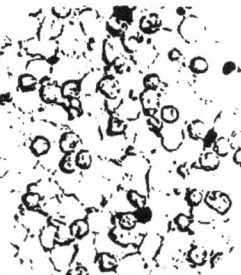

109.
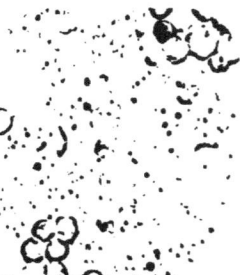

110.

111.
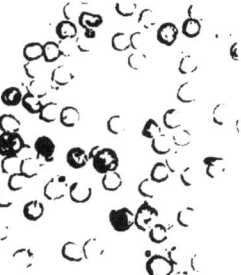

112.
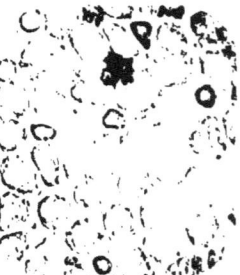

PLATE XV.

Photograph 101.—Primary, or least-developed lymph discs. They are about the diameter of the red corpuscles, and have the appearance of naked nuclei, but are in reality true cells, the walls of which lie in close contact with the nuclei.

Photograph 102.—Corpuscles from a fresh lymphatic gland of a pig, a little compressed by the cover-glass to show that the appearance seen in 101 is due to furrows or irregular depressions, and not to granules. The larger ones are more advanced, and have spread themselves down on the slide with great increase of diameter.

Photograph 103.—Appearances presented by lymph corpuscles taken from the same gland as 102, but from which the liquor lymphæ has been removed by the " method of isolation." Under these circumstances the primary ones show a dark edge, while the advanced ones have spread themselves down as white masses upon the slide.

Photograph 104 shows the peculiar sinking down and spreading tendency to which advanced lymph discs are liable after removal from their normal habitat.

Photograph 105.—Excessive action of sinking down and spreading, which comes on rapidly in the advanced lymph corpuscles when submitted to examination. The corpuscles which have still retained their form are of the primary kind.

Photograph 106.—Earliest appearance of spreading action in advanced lymph corpuscles. No reagents were added to the first six specimens on this plate.

Photograph 107.—A specimen of lymph diluted with fresh blister fluid. The advanced corpuscles have spread excessively, while the primary ones have retained their normal size.

Photograph 108.—In this case the spreading action of the advanced lymph disc has been restrained by diluting the specimen with a five per cent. solution of sodium chloride. The difference between the two kinds of corpuscles is still, however, sufficiently obvious.

PLATE XV.—*(continued.)*

Photograph 109.—Light and dark granules, the *debris* of both primary and advanced lymph corpuscles.

Photograph 110.—Primary and advanced lymph corpuscles from the thyroid gland of the calf. The advanced corpuscles are quite white, and spread down, while the primary ones are highly pigmented.

Photograph 111.—Corpuscles from the same thyroid gland as 110, preserved from spreading down by means of cane sugar. The distinctive appearance between the pigmented primary ones and the colourless ones is still sufficiently obvious.

Photograph 112.—Effects of staining lymph with saline aniline blue. The advanced corpuscles alone take the stain and swell up into enormous *discs*, while the primary ones become more or less contracted and shrunken.

PLATE XVI.   *Lymph and Splenic Discs, Action of Reagents.*

113.
114.
115.

116.
117.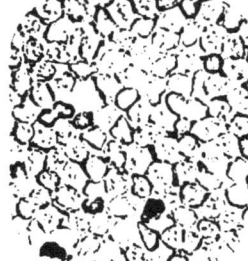
118.

119.
120.
121.

122.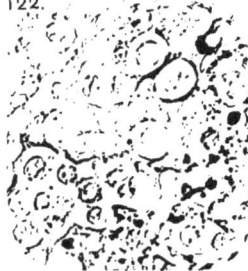
123.
124.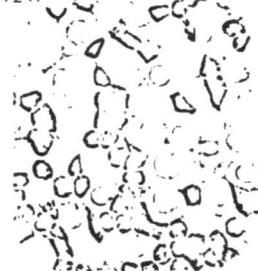

PLATE XVI.

Photograph 113.—Effects of saline aniline blue upon the corpuscles of sheep's lymph. Certain corpuscles stain and swell in proportion to their degree of development. These appear to be the free nuclei or advanced lymph corpuscles, inasmuch as the primary corpuscles, which are pigmented, neither stain nor swell, but, on the contrary, contract. Some corpuscles swell. but do not stain. This seems to indicate that the nuclei after their liberation undergo a further process of development, those which stain and swell most being the most advanced. These advanced corpuscles are very delicate and smooth, but being adhesive pick up granular matter which is free in the liquor lymphæ.

Photographs 114 and 115 are the same specimen before and after the addition of saline aniline blue, and are introduced to show the enormous degree of swelling which occurs under the influence of this reagent in both the corpuscles and the granular *debris.*

Photograph 116.—The advanced corpuscles in this specimen have been stained and swollen into large discs by the saline aniline blue. They are, however, of unequal diameter, and this denotes stages in their development. The primary pigmented corpuscles have, on the contrary, undergone contraction.

Photograph 117.—Corpuscles from the lymphatic gland of a pig. In this specimen three states of corpuscles were distinguishable after the addition of saline aniline blue. 1st, The primary pigmented kind, which did not stain but became contracted. 2nd, A kind which spread out into larger discs, but did not become stained. 3rd, A kind which became stained and extended into large discs of various diameters. These latter show in the photograph a granular or flocculent appearance.

Photograph 118.—Sometimes the granular *debris* which is attached to swollen and stained nuclei appears to be portions of the imperfectly removed primary capsule. In this case the granules do not stain. This appeared to be the case in this specimen.

Photograph 119.—Appearance presented by primary lymph corpuscles when spread down upon the slide and out of the liquor lymphæ. They are smooth, green, lustrous bodies, and may often be seen in bubble spaces. The white patches on this specimen are similar bodies attached to the cover. but out of focus, and must be neglected.

PLATE XVI.—*(continued.)*

Photograph 120.—A specimen of lymph from a lymphatic gland, in which an attempt has been made to preserve the advanced corpuscles by means of cane-sugar. They have, however, become partially laid down upon the glass-slide. The dark irregular bodies are corpuscles of the primary kind.

Photograph 121.—A specimen treated in the same manner as 120. The corpuscles are more perfectly preserved, but some of them are fusing together and spreading down upon the slide.

Photographs 122, 123, and 124 illustrate the endosmotic effects which can be induced in corpuscles of the primary order, and by which their true cellular character can be shown. By the passage inwards of liquid the capsule, which is usually closely applied upon the nucleus, becomes distended, and its presence demonstrated. 122 are primary corpuscles from the spleen of the pig, 123 from its lymphatic glands, and 124 from the thyroid of the calf.

PLATE XVII. *Comparison of Lymph and White Corpuscles.*

125.

126.

127.

128.

129.

130.

131.

132.

133.

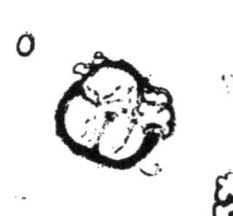

134.

135.

136.

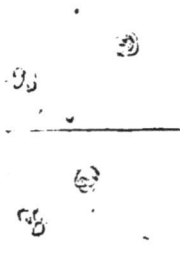

PLATE XVII.

Photograph 125.—White blood corpuscles in their usual spherical state in the absence of amœboid motion. They are somewhat stained with hæmoglobin. The red corpuscles have been washed away by ¾ per cent. salt solution. (Compare with Photograph 2, Plate II., in which they are less magnified, but are whiter and more irregular in form.)

Photograph 126.—Primary lymph corpuscles from the spleen of the pig, swollen and enlarged by being brought into contact with more limpid animal fluids than those in which they are usually found (liquor lymphæ). The nuclei swell up in such cases and continue to fill up the body of the cell. The action differs therefore from that depicted in Photographs 122, 123, and 124, Plate XVI., in which the cell wall simply is distended.

Photograph 127.—The same white corpuscles as seen in 125, reduced in size and approximated in appearance to primary lymph corpuscles by diluting their normal plasma with ¼ per cent. solution of sodium chloride. The clear edge has given place to a thick, dark outline.

Photograph 128.—Uninuclear stage of the white blood corpuscle, showing large nucleus stained with phosphene, which photographs of a dark colour.

Photograph 129.—Multinuclear white corpuscle in process of formation by division of the nucleus of the uninuclear variety. The faint bodies with dark outlines are decolourised red discs.

Photograph 130.—Example of binuclear white corpuscle, with nuclei fully developed, or nearly so, stained with phosphene.

Photograph 131.—Unstained examples of multinuclear white corpuscles, to which red corpuscles are adhering. They contain three or four nuclei.

Photograph 132.—Example of a white corpuscle, the capsule of which, and great part of the protoplasm, has become dissipated and the nuclei greatly swollen under the action of saline aniline blue. Scarcely anything is now left but four greatly extended nuclei.

Photograph 133.—Three white corpuscles adhering to each other, and also to modified colourless discs. On the surface of these corpuscles a crumpled pellicle or membrane is visible, which in one case is ruptured and partially withdrawn, exposing the white homogenous protoplasm of the interior.

PLATE XVII.—*(continued)*.

Photograph 134.—White corpuscles, which are spread down, but the pellicle of which has not been removed. On account of the spreading the nuclei are seen faintly through the envelope.

Photograph 135.—White corpuscles and their nuclei ruptured, emptied, and contracted by means of saturated solution of sodium chloride. The bodies in the background are decolourised red discs.

Photograph 136.—Protoplasm of the white corpuscle in a granular state stained with aniline scarlet. The nuclei have remained unstained.

PLATE XVIII.   *Cause of Biconcavity, Annulation, etc.*

137.

138.

139.
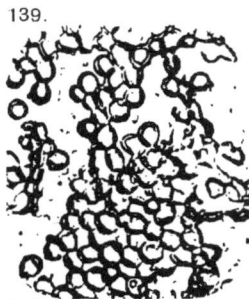

140.
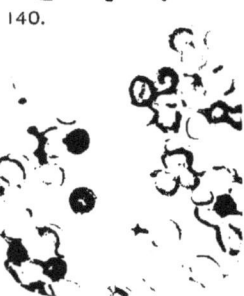

141.
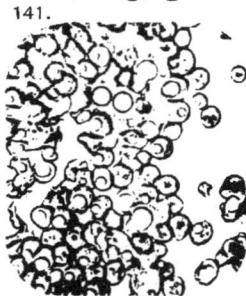

142.

143.
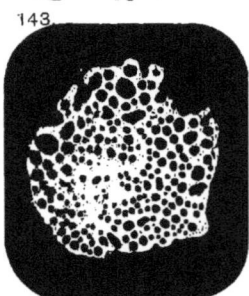

144.

145.
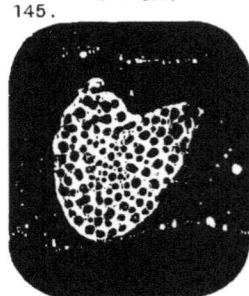

146.

147.

148.

PLATE XVIII.

Photograph 137.—Masses formed by the fusion together of colourless discs, fixed with osmic acid solution, and subsequently treated with water. They absorb the water, expand and annulate, *i.e.*, the substance is drawn to the edge, which becomes raised and thickened. Stained with aniline brown. The dark bodies in the interior are adhering red discs.

Photograph 138.—The extreme operation of the effect seen in progress in 137. The masses have been converted into perfect rings of regular and irregular shapes.

Photograph 139.—Red corpuscles, slightly fixed with weak tannic acid, undergoing conversion into rings, and which, being adherent at their circumference to each other, give the appearance of chain-mail.

Photograph 140.—Young colourless and partly-coloured discs undergoing annulation after being influenced by osmic acid in solution.

Photograph 141.—Red corpuscles lying side by side and becoming fused together, and their natural biconcavities becoming deepened till complete perforation occurs, and circular holes are produced in the fused mass.

Photograph 142.—Fibres resulting from the breaking of rings, such as are seen in 138.

Photographs 143 and 145.—Appearances produced by annulation when a minute portion of fat or oil is placed upon the surface of clean water.

Photograph 144.—Myelin rings, produced by spreading myelin upon a glass-slide in a very thin film, and, after placing over it a cover-glass, running in ¾ per cent. solution of sodium chloride.

Photograph 146.—Myelin rings and biconcave discs, formed in the same manner as in 144. These products may be compared with the altered forms of the blood corpuscles seen in 138 and 140.

Photograph 147.—Myelin forms, which contain among them shapes which are comparable with the blood shapes seen in Photograph 148.

PLATE XIX.

Photograph 149.—Colourless, clear-bordered, nucleated cells of embryo blood, lying among red discs and opaque, granular, white corpuscles.

Photographs 150 and 151.—Nucleated corpuscles from embryo blood, showing all shades of colour between the colourless ones seen in 149 and the full red nucleated corpuscle ; also the gradual diminution of size of the latter, and apparent atrophy of their nuclei in the process of red disc formation from red nucleated cells.

Photograph 152.—This specimen shows the clear-bordered, colourless, nucleated corpuscles of the blood of the embryo in process of development from a body in all respects similar to a primary lymph corpuscle, one of the sources of which seems to be the spleen of the embryo.

Photograph 153.—Colourless, clear-bordered, nucleated cells in process of development in the spleen pulp of the embryo. The dark corpuscles are contaminations from the blood. In some of these red corpuscles a trace of the nucleus is still apparent.

Photograph 154.—Bone-marrow from the rib of human embryo, showing nucleated corpuscles in all stages of development and colour, from the clear-bordered colourless to the full red ones, also their diminution of size, and apparent disappearance of nuclei prior to conversion into red discs.

Photograph 155.—Bone-marrow from another rib of the same human embryo as 154, showing the *new mode* by which red discs are formed at a late period in the embryo bone-marrow by the *direct colouration of nuclear bodies*, which are the analogue of the *colourless discs* of adult mammal blood instead of by changes induced in red nucleated corpuscles.

Photograph 156.—Coarsely granular pigment cells of the embryo bone-marrow. These corpuscles are undergoing regressive change : the coarse granular protoplasm is being removed, and a lighter-coloured, finely-granulous interior body is coming into view. The dark masses are red corpuscles.

Photograph 157.—The interior *nucleus* which is set free in the embryo bone-marrow by the dissipation of the outer envelope and of the coarse, dark, granular protoplasm of the primary pigment cell.

PLATE XIX.   *Embryo Blood, Splenic, and Bone-marrow Corpuscles.*

149.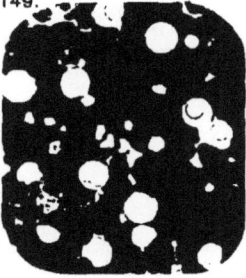

150.

151.

152.

153.

154.

155.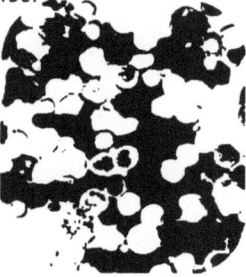

156.

157.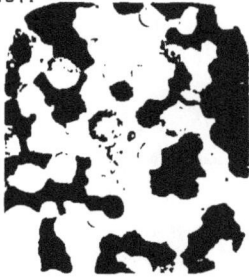

158.

159.

160.

PLATE XIX.—*(continued).*

Photographs 158 and 159.—Corpuscles from the bone-marrow of the sternum of a full-grown adult guinea-pig. These photographs show the entire series of transitions from the coarse, dark, granular pigment cell to the fully-developed red nucleated corpuscle. The dark, granular protoplasm disappears and reveals a cell which gradually attains colour, and becomes the red nucleated corpuscle. This in its turn undergoes changes which convert it *bodily* into a red disc.

Photograph 160.—Red corpuscles from the bone-marrow of the adult guinea-pig, which simulate coloured nuclei. They adhere together in groups, and also attach themselves readily to the ordinary blood discs present. They are either cells, the nuclei of which are obscured by the colouring matter or free nuclei, like those seen attaining colour in the bone-marrow of the embryo at the later periods of intra-uterine life. If the latter view is correct, the adult guinea-pig represents in its bone-marrow all the processes of blood production we obtain in the embryo of the adult mammal.

PLATE XX.
Diagram I.—Mammalian Blood.
MAJOR MODE.

Fig.   1. Primary lymph, splenic, and bone-marrow corpuscle.
Fig.   2. Advanced lymph, splenic, and bone-marrow corpuscle.
Fig.   3. Colourless disc of the blood.
Figs. 4, 5, and 6. Blood discs in various stages of colour.
Fig.   7. Full red blood disc.

MINOR MODE.

Fig.   1. Primary lymph, splenic, and bone-marrow corpuscle.
Fig.   2′. Uninuclear white blood corpuscle.
Fig.   3′. Uninuclear white corpuscle increased in size ; nucleus beginning to divide.
Fig.   4′. Further stage in division of nucleus.
Fig.   5′. Complete division of nucleus.
Fig.   6′. Nucleus undergoing division into three.
Fig.   7′. Nucleus undergoing segmentation into four.
Fig.   8′. Complete segmentation of nucleus into four.
Fig.   9′. Liberated growing nuclei, the equivalents of the advanced lymph discs. These are *invisible* in the blood till they have acquired a tinge of colour in advance of the liquor sanguinis.
Fig. 10′. Colourless biconcave disc.
Figs. 11′, 12′, and 13′. Intermediately coloured blood discs.
Fig. 14′. Full red disc.

*Diagram II.—Oviparous Blood.**
MAJOR MODE.

Fig.   1. Primary lymph, splenic, and bone-marrow corpuscle.
Figs. 2, 3, and 4. Clear-bordered, colourless, nucleated ellipsoids in various stages of growth, as seen both in the blood, and in the organs of corpuscular supply (spleen, bone-marrow, etc.)
Figs. 5, 6, and 7. Nucleated ellipsoids obtaining colour in the blood.
Fig.   8. Full red nucleated blood corpuscle.

MINOR MODE.

Fig.   1. Primary lymph, splenic, and bone-marrow corpuscle.
Fig.   2′. Uninuclear white blood corpuscle.
Fig.   3′. Nucleus beginning to divide.
Figs. 4′ and 5′. Further stages of segmentation, multinuclear corpuscles.
Fig.   6′. Liberated nuclei (nucleolated nuclei).
Fig.   7′. One of the free nuclei somewhat grown, and displaying a double membrane.
Figs. 8′ and 9′. More developed state ; nucleus and cell body becoming elliptical.
Figs. 10′ and 11′. Forms more developed, and becoming coloured.
Fig. 12′. Full red nucleated corpuscle.

* All the bodies sketched in this diagram may be seen photographed on Plate XXIII.

PLATE XX. *Development of Mammalian and Oviparous Blood.*

DEVELOPMENT OF MAMMALIAN BLOOD.

Diagram I.

MAJOR MODE.

MINOR MODE.

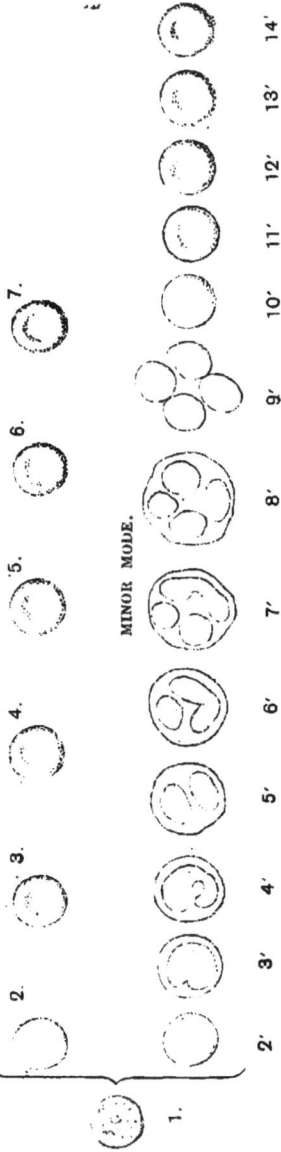

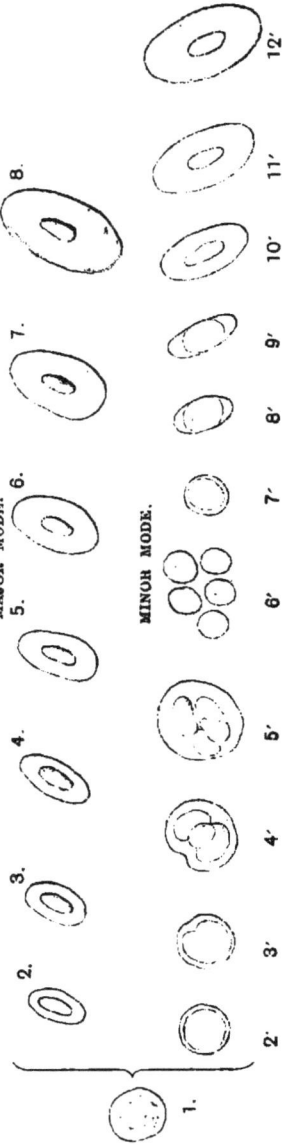

DEVELOPMENT OF OVIPAROUS BLOOD.

Diagram II.

MAJOR MODE.

MINOR MODE.

PLATE XXI.   *Adult Mammal Bone-marrow Corpuscles.*

161.

162.

163.

164.

165.

166.

167.

168.

169.
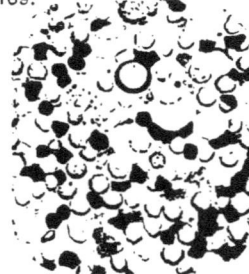

170.

171.

172.

PLATE XXI.

Photographs 161, 162, and 163 are corpuscles derived from the bone-marrow of the rib of the ox, and represent its lymphoid elements. In these elements we see the granular pigmentary matter clearing off, and exposing a second comparatively smooth body, which also contains a central mass (nucleolus or nucleus, according as to whether we view the cell as it stood originally, or after the removal of the primary envelope and contained protoplasm). This nucleolus, or nucleus, is in its turn set free, and is the body which leaves the bone-marrow and enters the blood as its colourless disc. It may sometimes attain a slight tinge of colour before leaving the marrow. The full red discs seen in these specimens are, of course, derived from the blood.

Photograph 164.—Corpuscles from the bone-marrow of the pig's rib. Some of the pigment corpuscles which take on the coarsely granular state are here seen in their smooth condition.

Photograph 165.—Products of the bone-marrow of the rabbit, showing the three kinds of corpuscles which spring from the typical pigment cell. The dark irregular masses are fused red corpuscles from the blood.

Photograph 166.—A specimen of bone-marrow of the ox, which shows very well the disintegration and peeling off of the pigmentary matter, and also the nucleated or nucleolated character of the body which is set free. Sometimes the nucleolus seems to be set free by a disintegration which involves both capsules simultaneously.

Photographs 167 and 168.—Examples in which the free nuclei which go to form the colourless discs of the blood are being set free by the simultaneous disintegration and stripping off of both capsules. The reactions of the saline aniline blue enable us to study these changes, inasmuch as when the specimens are fresh it stains nuclei or nucleoli only, leaving the capsular matter unaffected.

Photograph 169.—This specimen of bone-marrow substance is introduced to show that the smooth white bodies set free by the disintegration of the capsule and protoplasm of the pigment cell are similar in their nature to the primary lymph corpuscle, as they can be made to show by osmosis a delicate cell-wall. (Compare with Photographs 122, 123, and 124, Plate XVI.)

PLATE XXI.—*(continued).*

Photograph 170.—Fresh bone-marrow of pig, to which weak saline aniline blue has been added. The nuclei already set free, stain, and come at once into view as smooth spreading bodies. (The blue colour takes as white in the photograph.)

Photograph 171.—A specimen of bone marrow from the rib of the ox, showing that the body which lies in the interior of the dark granulating corpuscle sometimes contains more than one nucleus or nucleolus.

Photograph 172.—Specimen of bone-marrow from the ox, showing how the smooth pigment cells which give rise to the dark, coarsely granulated ones originate in the marrow by subdivision.

PLATE XXII. *Bone-marrow and Splenic Oviparous Corpuscles.*

173.

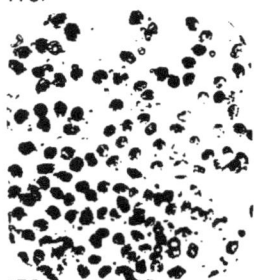

174.

175.

176.

177.

178.

179.

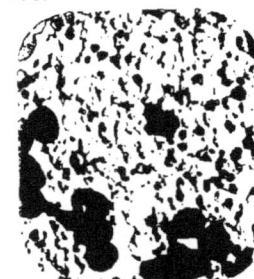

180.

181.

182.

183.

184.

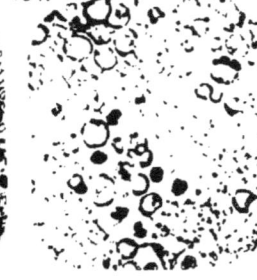

PLATE XXII.

Photograph 173.—Specimen from the bone-marrow of the common fowl, showing *colourless nucleated ellipsoids.* The clear margins are obliterated, confused, and indistinct.

Photograph 174.— Bone-marrow of fowl. The clear margins of the ellipsoids have been preserved with osmic acid solution, and the nuclei and extreme edges of the cells, which are now slightly granular, have been stained with fuschine.

Photograph 175.—Bone-marrow of fowl, showing the tendency which the nuclei of these ellipsoids have to swell and fill up the margins. This action is seen more excessively in Photograph 176.

Photograph 177.—A specimen from the fowl's bone-marrow, which, in addition to nucleated ellipsoids show the granular corpuscles from which they are derived. They are mostly fused together. Specimens of the same corpuscles may be seen on Photograph 174.

Photograph 178.—Coarsely granular pigment cell of the bone-marrow of the fowl, which liberates the nucleolated nucleus, from which the nucleated clear margined ellipsoid is produced. The specimen is contaminated with ordinary red corpuscles.

Photograph 179.—Spleen pulp of the pike, showing what appear to be simply a mass of nuclei, but which are in reality colourless, clear-margined ellipsoids. The red corpuscles must be neglected.

Photographs 180 and 181.—Specimens showing the nucleated colourless ellipsoids, which are the final product of the spleen of the pike. 180 was stained with a weak aqueous solution of aniline brown, and 181 by a stronger solution. The dark indistinct mass in 181 is a mass of these ellipsoids, the nuclei of which alone are visible.

Photograph 182.—Colourless nucleated ellipsoids from the spleen of the perch, stained with saline aniline blue, which has caused the nuclei to swell a little.

Photograph 183.—Colourless nucleated ellipsoids from the spleen of the perch, the clear margins of which are melting down and fusing, and allowing the nuclei also to come into contact, and to run together.

Photograph 184.—This specimen shows a few of the lymphoid cells from which the colourless, clear-margined, nucleated ellipsoids are developed. The smaller dark bodies are the nuclei of ellipsoids, the margins of which are obscured.

PLATE XXIII.

Photograph 185.—Ordinary white corpuscle of the frog's blood, in the opaque, furrowed, or corrugated condition. The red cells must be neglected.

Photograph 186.—The same body as 185, in a transparent state, in which its uninuclear character is revealed. The nucleus shows a slight indentation, which is the first step in the process of segmentation. The specimen shows also a free nucleus.

Photograph 187.—This specimen shows a corpuscle of the same character as the one seen in 186, and another in which the segmentation of the nucleus is so far advanced that three complete nuclei are already formed, and others are in process of formation.

Photograph 188.—The specimen shows a multinuclear corpuscle in an advanced state of development. The nuclei have increased much in size, and the body of the cell is disappearing. A grown free nucleus is also present.

Photograph 189.—Growing nuclei, which have been set free from a white corpuscle, and the exterior or capsules of which are becoming thickened prior to separating into two distinct layers.

Photographs 190, 191, 192.—These specimens show how the cellular body is developed about the nucleus by the formation or separation of an external wall. In most cases the nucleus becomes ovoid, and the cell then takes on an elliptical form. Some of these corpuscles get colour before they are much grown.

Photograph 193.—Bodies which look like naked nuclei from the blood of the Triton, but which are in reality large elliptical corpuscles, having colourless clear margins, which are invisible in the serum. These bodies are derived in the state in which they appear in the blood from the spleen and bone-marrow.

Photographs 194 and 195.—These corpuscles are the same as those described in 193, but they have been stained with fuschine, which has dyed the nuclei deeply and the clear margins just sufficiently to enable them to be photographed. The two bodies, with crumpled capsules showing a radiating appearance around the nucleus, are ordinary red corpuscles, which have been decolourised by the staining fluid. The difference between these and the corpuscles whose cellular bodies are, under ordinary circumstances invisible, is sufficiently obvious. The latter kind of corpuscle exists in all ovipara, and represents in them the major process of blood-making. It is the *analogue* of the colourless disc of the mammal blood.

Photograph 196.—Micrometer scale divided into 10 and 20,000ths of an inch, magnified like the objects in the specimens, 500 diameters.

185.

186.

187.

188.

189.

190.

191.
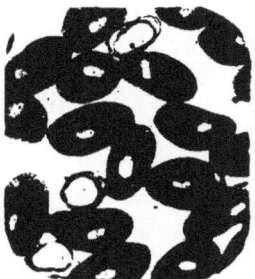

192
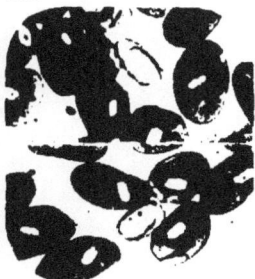

193.

194.

195.
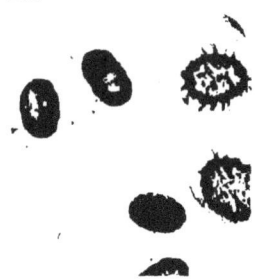

196

On the 14th of January, 1882, a paper by Professor
Bizzozero, of Turin, bearing the title "Ueber einen neuen.
Formbestandteil des Säugetierblutes, und die Bedeutung
desselben für die Thrombosis und Blutgerinnung Ueberhaupt,"
was published in the "Centralblatt für die Medicinischen
Wissenschaften" as an original communication.

This was followed by a leader in the "Lancet" of
January 21st, headed "A New Blood-Corpuscle," in which the
claims of Bizzozero were brought prominently before the English
scientific public. This leader called forth two letters in the
next issue of the "Lancet" (Jan. 23, 1882), one from Dr. Richard
Neale, and one from myself. On February 9th, at a meeting
of the Birmingham Philosophical Society, Dr. Heslop drew the
attention of the members of the Society to the singular coinci-
dences in the statements made by Professor Bizzozero to
those which were contained in the author's papers published
in the Transactions of the Society in 1878 and 1880, the
years of Dr. Heslop's presidency. At this meeting a resolu-
tion was passed, ordering Dr. Heslop's abstract to be
placed upon the minutes of the Society, and requesting the
Council to transmit a copy of it to the President and Council of
the Royal Society and to the leading scientific periodicals.
This was subsequently carried out. In the meantime Professor
Bizzozero had published a further communication in the issue
of the "Centralblatt" of March 10th, bearing the title
"Die Blutplättchen der Säugetiere und die 'Invisible Corpuscles,' von Norris," in which he replied to the letters of
Dr. Richard Neale and myself in the "Lancet" of January
23, 1882.

On March 18th a second leader appeared in the
"Lancet," under the heading of "Norris and Bizzozero," in
which the abstract received from the Philosophical Society was
acknowledged, and the various letters on the subject adverted to

As the properties which Bizzozero ascribes to the corpuscle
he has seen are precisely those which long ago I published as
being possessed by the corpuscles which I have designated the
fugitive group of discs, it follows that either Bizzozero's corpuscle
belongs to this series, or that it is an independent one, having
the same properties as these possess. To determine this point
I have examined the blood under the conditions present in
Bizzozero's experiments, and have communicated my conclusions to the "Lancet" in a paper published in its issue of
April 8th, 1882, which will be found on page 250 of this
Appendix.

The reader is now in possession of the *whole* of the evidence,
and to his unbiassed judgment the decision must be left.

ÜEBER EINEN NEUEN FORMBESTANDTEIL DES SÄUGETIERBLUTES, UND DIE BEDEUTUNG DESSELBEN FÜR DIE THROMBOSIS UND BLUTGERINNUNG ÜBERHAUPT.

VORLÄUFIGE MITTEILUNG VON. PROF. G. BIZZOZERO IN TURIN.

" Beobachtet man unter starker Vergröfserung den Blut-
kreislauf in den kleinen Gefäfsen der Säugetiere (im Mesen-
terium chloralisirter Kaninchen oder Meerschweinchen), so
gelangt man bald zu dem unerwarteten Ergebnisse, dass darin,
aufser den roten und farblosen Blutkörperchen, noch ein drittes
Formelement circulirt. Dasselbe wird dargestellt durch sehr
blasse, farblose, ovale oder runde, scheiben- oder linsenförmige
Plättchen, von drei- bis zweimal geringerem Durchmesser als
die roten Blutkörperchen und regellos unter diesen zerstreut
circulirend. — Wenn diese Gebilde den vielen Forschern, die
sich mit der Beobachtung des circulirenden Blutes beschäftigt
haben, entgangen sind, so scheint dies an mehrfachen
Umständen zu liegen, und zwar : a) an der Farblosigkeit und
Durchsichtigkeit der Plättchen : b) daran, dass sie viel spär-
licher sind als die roten, und minder sichtbar, als die weifsen
Blutkörperchen, weshalb nur die beiden letzteren Formen die
Aufmerksamkeit des unvorbereiteten Beobachters auf sich
ziehen ; c) auch wohl an dem Umstande, dass bei den
Säugetieren die directe Beobachtung der Blutströmung in den

kleinen Gefäfsen mit gröfseren Schwierigkeiten verbunden ist und daher solche Untersuchungen meistens nur an Kaltblütern vorgenommen wurden.

" Die Plättchen sind auch in ganz frisch entzogenem Blute sichtbar ; sie erscheinen grofsenteils gehäuft um die farblosen Blutkörperchen, oder steigen in die oberen Schichten der Flüssigkeit auf, wo sie sich an das Deckgläschen anlegen. — Diese Plättchen sind es, die in dem frisch entzogenen Blute durch ihre rasche Alteration und Verunstaltung ein körniges Aussehen gewinnen und so jene Körnchenhaufen erzeugen, die bereits von so vielen Histologen im Blute beschrieben worden sind.

"Durch geeignete Reagentien kann aber die Form der Plättchen unverändert erhalten und diese letzteren selbst einer längeren Beobachtung zugänglich gemacht werden ; z. B. mittels einer durch Methylviolet gefärbten indifferenten Kochsalzlösung, welche auch den Vorzug gewährt, sowohl die roten, als auch die farblosen Blutkörperchen wohl zu erhalten. Beim Menschen, wo die Blutplättchen sich sehr leicht alteriren, empfiehlt sich zu ihrer Untersuchung folgendes Verfahren : Man sticht einen Finger an, bringt auf die Stichwunde einen Tropfen von der obigen Kochsalzlösung und lässt dann durch Zusammendrücken des Fingers ein Tröpfchen Blut hervorquellen, das nun unmittelbar in Berührung tritt mit der Lösung und nach gehöriger Durchmischung mit derselben zur mikroskopischen Untersuchung verwendet wird. — Die schönsten Plättchen erhält man vom Meerschweinchen.

" Zur Zeit bin ich nicht in der Lage, etwas Positives über die Herkunft der Blutplättchen auszusagen. Jedenfalls aber spricht Nichts für die Ableitung derselben von dem Zerfalle der farblosen Blutkörperchen ; denn die Plättchen besitzen eine typische Form, und im Inhalte der farblosen Blutzellen finden wir keinen Bestandteil, der ihnen einigermafsen gleicht.

" Der Vergleich zwischen dem entzogenen und dem circulirenden Blute löst die bisher offen gebliebene Frage von den sogenannten Körnchenhaufen des Blutes, die von den meisten Autoren als wahre, von dem Zerfalle der farblosen Blutkörperchen herrührende Körnchen aufgefasst werden, während Andere (wie z. B. HAYEM) sie von der Umwandlung eigener, im Blute präformirt enthaltener Plättchen ableiten. — Nun handelt es sich in der Tat um Plättchen. — Doch hat HAYEM in Betreff ihrer Präexistenz nur eine Hypothese ausgesprochen, da er nicht das circulirende Blut der Säugetiere untersucht hat. Auch beschrieb und deutete er die Plättchen irrtümlich, indem er sie als biconcave Scheibchen schilderte und für Elemente hielt, die zu roten Blutkörperchen sich zu verwandeln bestimmt wären, weshalb er sie mit dem Namen Hämatoblasten belegte.

Indessen bestehen die besagten Gebilde aus einer vom Stroma
der roten Blutkörperchen sehr verschiedenen Substanz und
enthalten niemals Hämoglobin.

"In Zukunft also wird man bei dem Studium der Functionen
und der Alterationen des Blutes stets auch diesen constanten und
reichlich vertretenen neuen Formbestanteil mit in Betracht
ziehen müssen. Die Wichtigkeit desselben erhellt mir schon
jetzt sowohl aus seiner numerischen Zunahme bei vielen krank-
haften Zuständen (wie z. B. nach dem Aderlasse), als aus den
Beobachtungen, die ich hierüber bei der Thrombenbildung und
bei der Gerinnung des Blutes gemacht habe.

" Was zunächst die Thrombosis anlangt, so bilden die Blutplättchen den überwiegenden Bestandteil des weifsen Thrombus
der Säugetiere, indem sie jene körnige Substanz abgeben, die
man zwischen den farblosen Blutkörperchen vorfindet, und
bisher vom Zerfalle dieser letzteren oder von der Gerinnung des
Faserstoffes abzuleiten pflegte.

"Hinsichtlich der Gerinnung des Blutes fällt wahrscheinlich
den Plättchen jene Rolle zu, welche MANTEGAZZA und A.
SCHMIDT den farblosen Blutkörperchen zuschreiben. — Schon
SCHULTZE, RANVIER, HAYEM u. A. hatten bemerkt, dass die
netzartig verbundenen Fibrinfäden in einem gerinnenden
Blutstropfen oft in den oben erwähnten Körnchenhaufen zusam-
menlaufen und schlossen daraus auf einen Zusammenhang
zwischen diesen Körnchen und der Füllung des Faserstoffes. HAYEM ging noch etwas weiter und fand, dass einige
Flüssigkeiten, welche die Blutgerinnung verzögern, die Form
seiner " Hämatoblasten " unverändert erhalten. — Nach A.
SCHMIDT wird die Gerinnung des Blutes durch die weissen
Blutkörperchen bedingt ; ja es sollen eben diese letzteren durch
ihren massenhaften Zerfall die erwähnten Körnchenhaufen
erzeugen, und so zu einem grofsen Teile das Material, woraus
der Faserstoff besteht, liefern.

Meinerseits bin ich durch folgende Gründe zu der Annahme
veranlasst, dass es nicht die farblosen Blutkörperchen, sondern
die Blutplättchen seien, die den Ausgangspunkt der Gerinnung
abgeben : 1) Habe ich mich nie von jenem massenhaften
Zerfalle der farblosen Blutkörperchen, wie er von A. SCHMIDT
angenommen wird, überzeugen können. Denn auch im circulirenden Blute der Säugetiere sind die weifsen Blutkörperchen
nur sehr spärlich vertreten, und bei Erzeugung einer Blutung
habe ich (vorausgesetzt, dass das ausfliefsende Blut in einer
indifferenten Flüssigkeit aufgefangen wurde) nie vermocht, unter
meinen Augen die farblosen Blutkörperchen zerfallen zu sehen.
Giebt es aber auch einen kleinen Unterschied in dem
numerischen Verhältnisse zwischen farblosen und farbigen
Blutkörperchen in- und ausserhalb der Gefäfse, so erklärt

sich derselbe daraus, dass die farbigen leichter, als die
farblosen durch die Gefäfswunde austreten. — 2) Wenn man
ein soeben entzogenes Bluttröpfchen beobachtet, so zeigt
sich, dass die Zeit, binnen welcher es darin zur Gerinnung
kommt, dem Zeitraume entspricht, innerhalb dessen die
Blutplättchen der Entartung anheimfallen. — Die Flüssigkeiten,
welche die Gerinnung verspäten oder verhindern (z. B. Lösungen
von kohlensaurem Natron oder schwefelsaurer Magnesia) verhindern auch, so weit ich bisher gesehen, die körnige Verwand-
lung der Blutplättchen. Eine indifferente Kochsalzlösung
bewahrt die Plättchen nicht, während es eine eben solche, aber
durch Methylviolett gefärbte Lösung wohl tut ; nun gerinnt
aber das Blut in ersterer binnen einer Viertelstunde, während
ich es in der methylvioletthaltigen Lösung nach 24 Stunden
noch ungeronnen fand. — 3) Fasst man eine Blutgefäfsstrecke
beim lebenden Tiere zwischen zwei Ligaturen, so bleibt das
darin enthaltene Blut stundenlang flüssig, und während dieser
ganzen Zeit bewahren die Blutplättchen darin ihre charak-
teristische Form ; während in dem dem Einflusse der Gefäfs-
wand entzogenen Blute des Aderlasses sie in weniger als einer
Minute der Entartung unterliegen. — Wenn man Blut mit
Zwirnfäden schlägt, aber bevor die Gerinnung eingetreten, die
Fäden herauszieht (beim Hundeblute, wenn es in 0,75 proc.
NaCl-Lösung aufgefangen wurde, muss dies z. B. schon nach
45 Secunden geschehen) und sie dann in eine, die Blutplättchen
conservirende Flüssigkeit eintaucht, so zeigt die mikroskopische
Untersuchung, dass nur sehr wenige farblose Blutkörperchen
an den einzelnen Fäserchen der Fäden haften geblieben sind ;
diese letzteren aber von einer dicken Schicht Blutplättchen
(welche viscös geworden und sich daher an den Zwirnfasern
festgeklebt haben) überzogen sind. — Dauert das Schlagen
länger, so entarten die Plättchen und bleiben in den Faserstoff-
schichten stecken.

Es ergiebt sich also, dass, während die weifsen Blutkörperchen beim Eintritte der Gerinnung keine merkliche Verän-
derung erleiden, die Blutplättchen dagegen sich dabei sehr
erheblich alteriren ; dass ferner der Faserstoff sich gerade dort
niederschlägt, wo sich die Plättchen anhängen ; dass endlich
die Mittel, welche der Entartung der Blutplättchen entgegen-
wirken, zugleich auch die Blutgerinnung verzögern oder
verhindern. Dies Alles zusammengenommen, macht es mehr
als wahrscheinlich, dass die Gerinnung des Blutes unter dem
directen Einflusse der Blutplättchen stehe.

A NEW BLOOD-CORPUSCLE.

From the " Lancet," January 21st, 1882.

The discovery of a new and important constituent of the mammalian blood has just been announced by a distinguished investigator of blood formation—Professor Bizzozero of Turin. This new element is not the same as the invisible corpuscle of Norris, but presents nevertheless somewhat similar characters. If the course of the circulation is watched in the small vessels in the mesentery of chloralised rabbits and guinea-pigs, there are seen, besides the ordinary red and pale corpuscles, third elements—very pale, oval, or round disc-shaped or lenticular bodies, one-half or one-third the diameter of the red corpuscles, among which they are scattered. " Blutplättchen," Bizzozero proposes to call them. They have hitherto escaped notice, probably because they are so colourless and translucent, less numerous than the red, and less visible than the white corpuscles ; and on account of the difficulty of observing the mammalian blood in the course of the circulation with a high magnifying power. They are to be observed also in freshly drawn blood, for the most part aggregated around the colourless corpuscles, or, ascending to the upper layer, they adhere to the cover-glass. They change, however, with great rapidity, rapidly become granular, and appear to be the source of the small granule masses which have been described by many observers. The corpuscles can be preserved unaltered in form for more prolonged examination by certain reagents, as, for instance, by a solution of chloride of sodium tinted with methyl-violet. They are to be found also in human blood, but they undergo alterations with extreme rapidity, and the best method of observing them has been found to be by placing a drop of the above solution over the puncture, and then squeezing the blood out, and immediately examining it under the microscope.

Bizzozero has been unable as yet to ascertain anything regarding the origin of these elements. It is exceedingly improbable that they are in any way derived from the ordinary colourless corpuscles, because they possess a very definite and characteristic form, and the leucocytes contain no element from which these objects could be derived. A comparison between the blood in the vessels and out of the body thus clears up the origin of the granule heaps, which some regard as products of the destruction of leucocytes, and others, as Hayem, ascribe to changes in peculiar flat corpuscles. The latter view is undoubtedly correct, although Hayem does not seem to have observed these elements in the circulating blood, since he describes them as biconcave discs which are transformed into red corpuscles, and calls them " hæmatoblasts." The objects regarded by Bizzozero as the source of the granules possess no stroma, and never contain hæmoglobin ; they differ therefore from the hæmatoblasts of Hayem.

The new elements seem to play an important part in the functional alterations of the blood. They are increased in certain morbid conditions—as, for instance, after bleeding—and play an important part in the production of thrombi. They constitute the chief part of the white clots in the mammalia, since they give rise to the granular material which is seen between the pale corpuscles, and which has hitherto been ascribed to the degeneration of fibrin. In the process of coagulation these elements appear to exert the influence which has been attributed by Mantegazza and Schmidt to the colourless cor- puscles. Schultz, Ranvier, Hayem, and others, have noted that the reticulated threads of fibrin often present at their junction these groups of granules, and hence inferred that the latter were produced by the degeneration of the fibrin. Hayem, however, found that certain fluids which hinder coagulation preserve unchanged the form of his " hæmatoblasts." It will also be remembered that A. Schmidt asserted that the coagula- tion of the blood is effected by the white corpuscles, which by their destruction yield the granules, and so constitute a con- siderable part of the substance of the clot. Bizzozero, however, now urges that the formation of the clot is due, not to the white corpuscles, but to these new elements. He has never been able to satisfy himself of the wholesale destruction of white corpuscles assumed by Schmidt. Leucocytes are comparatively few in the circulating blood, and he could never observe any destruction of them after the blood was drawn, provided it was mixed with an indifferent fluid, such as a saline solution. The time at which coagulation occurs in a given drop of blood corre- sponds closely to that at which these new elements present the degenerative changes. The fluids which retard or prevent coagulation—solutions of carbonate of soda or of sulphate of magnesia, for instance—also hinder the granular transformation of the new corpuscles. The indifferent solution of chloride of sodium does not preserve them, but one to which methyl-violet has been added does so. With the former the blood coagulates in a quarter of an hour, with the latter it remains liquid for twenty-four hours. If a vessel of a living animal is included between two ligatures, the blood within it remains liquid for hours, and during the whole time these elements preserve their characteristic form, although in blood outside the vessels they undergo degeneration in a few minutes. If blood is " whipped"

and the fibres employed are withdrawn before coagulation commences, and are then immersed in a liquid capable of preserving
the new elements unaltered, it will be found that they are
covered with a thick layer of the new elements, among which
are very few white corpuscles. If the whipping has been continued longer, these elements are found to have undergone
degeneration and to remain on the layer of fibrin. From these
facts it follows that whereas the ordinary white blood corpuscles
present no noteworthy changes at the commencement of coagulation, these new elements are considerably altered, and where
they adhere, there the fibrin is deposited, and, finally, that all
agents which hinder their transformation retard also the coagulation of the blood. The evidence is thus very strong that this
coagulation—that is, the formation of fibrin—takes place under
the direct influence of these corpuscles.

THE NEW BLOOD-CORPUSCLE.

To the EDITOR *of* THE LANCET.

SIR,—In a leader of your issue of the 21st instant you
bring before the notice of your readers certain recent observations made on the blood by that distinguished investigator,
Professor Bizzozero, of Turin. This author has seen in the
circulating blood in the mesentery of the rabbit and the guinea-
pig certain third elements, which are neither the white nor the
red corpuscles, and which he describes as " very pale, oval, or
round, disc-shaped or lenticular bodies, one-half or one-third
the diameter of the red corpuscles among which they are
scattered." These, it is said, have hitherto escaped notice,
" probably because they are so colourless and translucent, and
less numerous than the red and less visible than the white
corpuscles." These bodies can also be observed in freshly drawn
blood; they aggregate around the white corpuscles, or,
ascending to the upper layer, adhere to the cover glass. They
have also a great tendency to change and become granular,
and give rise to the granular masses which have been described
by many observers. These corpuscles can be more or less
perfectly preserved by a solution of sodium chloride, tinted
with methyl-violet, and the best method of observing them has
been found to be by placing a drop of the solution over the
puncture and squeezing the blood into it and examining
immediately under the microscope. Bizzozero has not ascer-

tained anything as to the origin of these elements. These bodies and the granules they give rise to are, however, considered to have an important relation to the formation of fibrin.

You say—"This new element is not the same as the invisible corpuscle of Norris, but presents, nevertheless, somewhat similar characters." This statement, though partially correct, is very misleading. While it is obvious that Bizzozero cannot have *seen* in the blood a corpuscle which is *invisible*, it must not be overlooked that my discovery is not simply that of an *invisible* corpuscle, but of a graduated series of corpuscles, the youngest members of which possess no hæmogoblin, and are, under normal conditions, invisible in the blood, while the oldest and most advanced possess just enough tint to make them contrast slightly with the liquor sanguinis, and therefore to be visible. This series of corpuscles I have designated the "fugitive group," to distinguish them from the ordinary red discs, which are more permanent, and also to indicate their characteristic tendency to undergo disintegration, and to form fibrin when the blood is shed. The barely visible corpuscles of this group are the "very pale, oval, round, disc-shaped or lenticular bodies" recently seen in the circulating blood by Bizzozero.

The title of my paper on the blood read to the Philosophical Society of Birmingham, on November 14th, 1878, runs as follows:—"On the Existence in Mammalian Blood of a Morphological Element, which explains the origin of the red disc and the formation of fibrin." After giving many methods by which the existence of this corpuscle can be demonstrated, I conclude as follows:—1. That there exist in the blood of mammalia, in addition to the well-known red and white corpuscles, colourless, transparent, biconcave discs of the same size as the red ones. 2. Between these two kinds of biconcave discs others are demonstrable, having every *intermediate* gradation of colour. To get at the real size and form of these corpuscles it was necessary to use various means of preservation, for, on account of their liquidity and extreme susceptibility to change, they often present themselves as small spheres, at other times as discs, and at others, liquid-like, take the shape of the interstices in which they happen to lie. They are also prone to separate readily into smaller portions, and to granulate and form masses. All this I have explained in detail in my papers. That I knew the power of sodium chloride to preserve these corpuscles is seen in the following:—"It is well known that the coagulation of the blood can be entirely prevented by means of saturated solutions of neutral salts. I have ascertained that this is due to the power of these substances to maintain the integrity of the *invisible and subcruorised corpuscles,* which is the *true fibrin stroma.*" In a footnote to this passage I make the following statement:—"The discovery of a third corpuscle has thrown great light on the question of the coagulation of the blood, and of fibrin formation generally. To avoid complication, this subject will receive separate treatment. It may, however, be briefly stated here that on the basis of their behaviour when the blood is shed the biconcave discs are divisible into two groups—a fugitive and a permanent group—and that the changes which take place in the former determine coagulation. It is not a little singular that Bizzozero follows the same mode of staining these corpuscles." I say—" Place upon the end of the finger a small drop of the staining fluid (saline-aniline-blue), and with a needle prick the finger through this drop, so that the blood may, when the finger is squeezed, flow directly into the liquid, which has the double property of preserving and staining." I had also observed the tendency of these corpuscles to adhere to the cover-glass, and, when speaking of their specific gravity, said:—" Like the white corpuscles, they are *lighter* than the red, and have a tendency constantly to rise to the surface of the blood, consequently the largest numbers are always seen to attach themselves to the *upper glass* in preference to the lower, especially if time be allowed them to rise. This, no doubt, has something to do with the *buffy coat.*"

Many more facts might be mentioned : but enough has been said to show that my discoveries, made in 1877 and published in 1878, cover the whole ground, and that the recent researches of Bizzozero cannot be regarded in any other light than as a most important and valuable confirmation of my views.

I am, Sir, yours faithfully,

RICHARD NORRIS, M.D., F.R.S.E.

Birmingham, Jan. 23rd, 1882.

—————

To the EDITOR *of* THE LANCET.

SIR,—In your journal of January 21st, p. 111, you refer to Professor Bizzozero's researches on the blood, and state that the " Blutplättchen " he has recently discovered are likely to lead to important results. In the *Medical Times and Gazette*, April, 1854, p. 430, you may find described a number of examinations of the blood of goitrous patients, and therein what appear to me to be bodies identical with those now brought under notice by

Professor Bizzozero. I remember thinking at the time the observations were worthy of notice, but they did not seem to strike anyone else in the same light, so that now, after a lapse of twenty-seven years, I am gratified by seeing them reintroduced under foreign patronage.

I am, Sir, yours, &c.,
RICHARD NEALE, M.D., Lond.

Boundary Road, South Hampstead, N.W.,
Jan. 23rd, 1882.

BIRMINGHAM PHILOSOPHICAL SOCIETY.

Note on the Papers by Dr. Norris *" On the Development of Mammalian Blood," published in the Proceedings of the Society (Vol. 1., 1878-1879, and Vol. 11., 1879-1880), in relation to the " Discovery" announced by Professor Bizzozero.*

Extracted from the Minutes of a Meeting of the Society held on February 9th, 1882.

Dr. Heslop, the President of the Society in the years 1878-9 and 1879-80, drew attention to a leading article in the *Lancet*, dated January 21, 1882, which opens with the following words :—" The discovery of a new and important constituent of the mammalian blood has just been announced by a distinguished investigator of blood formation—Professor Bizzozero, of Turin. This new element is said not to be the same as the invisible corpuscle of Norris, but presents, nevertheless, somewhat similar characters." Further on, these third elements are spoken of as " very pale, oval, or round disc-shaped or lenticular bodies, one-half or one-third the diameter of the red corpuscles, among which they are scattered." Further details are given as to the part taken by the elements described by Bizzozero in the formation of granule-heaps, in the production of thrombi, and in the coagulation of the blood. The formation of the clot is declared to be due not to the white corpuscles, but to these new elements, which undergo alterations with extreme rapidity. Finally, it is averred that the best mode of observing them in freshly drawn blood is by placing a drop of a solution of sodium chloride tinted with methyl-violet over a puncture. The blood being then squeezed out should be immediately examined under the microscope.

The statements made in the *Lancet* seem to be based on a paper communicated by Professor Bizzozero, of Turin, to the " Centralblatt für die Medicinischen Wissenschaften," dated 14th January, 1882, and accurately reproduce in abstract the leading facts related in that paper. The allusion, however, to Dr. Norris is only to be found in the *Lancet*, his name and researches being alike unmentioned by the Turin Professor. It hence appears needful to draw attention to the dates and subject-matter of Dr. Norris's remarkable papers on the blood in the " Transactions of the Birmingham Philosophical Society."

The date of Dr. Norris's first paper was November 14, 1878. Its title was, " On the existence in mammalian blood of a new morphological element which explains the origin of the red disc and the formation of fibrin." It is divided into three parts, and is copiously illustrated by photographs. In the first paper he alludes to the value of photography as an instrument of research, and speaks of its having " detected the existence of corpuscles which differed so little in refractive power and colour from the liquor sanguinis as to be invisible to the eye." He was then led to form the opinion that " possibly other corpuscles might exist, having precisely the same refractive index and actinic value as the liquor sanguinis, and that such would not only be invisible, but also incapable of being photographed." He then proceeds to describe in minute detail his methods of examination, and draws attention to the difficulties encountered in the attempt to make these corpuscles visible by means of stains. 1. " These corpuscles are extremely fugitive in their character when the blood is shed, and are rendered more so by dilution of the plasma. 2. Like the red discs, they stain with great difficulty, so long as they remain submerged in the liquor sanguinis, or in saline solutions. 3. Substances which stain these corpuscles also give the same tint to the liquor sanguinis, and, therefore, the former still remain obscured." After alluding to the well-known power of saturated solutions of neutral salts to prevent the coagulation of the blood he says : " I have ascertained that this is due to the power of these substances to maintain the integrity of the invisible and sub-cruorised corpuscle, which is the true fibrin stroma." In a note to this passage (Proceedings, vol. i., part 2, page 13,) the following words occur :—" The discovery of a third corpuscle has thrown great light on the question of the coagulation of the blood, and of fibrin formation generally." " It may, however, be briefly stated here on the basis of their behaviour when the blood is shed, that the biconcave discs are divisible into two groups—a fugitive and a permanent group, and that the changes which take place in the former determine coagulation." At the end of this, the first

part of his paper, the author draws these two conclusions :—
1. That there exists in the blood of mammalia, in addition
to the well-known red and white corpuscles, colourless transparent biconcave discs of the same size as the red ones. 2·
Between these two kinds of biconcave discs others having
every *intermediate* gradation of colour are demonstrable."

The title of the second part of the paper is, " On
the Origin of the Colourless Biconcave Disc of Mammalian
Blood." He first describes the characters of the lymph
and splenic corpuscles, and alludes to the capacity of
these discs to undergo conversion into smooth colourless
biconcave discs, and considers that this " seems to point
definitely to the *source* of the colourless biconcave discs found
in the blood." Further on he shows the differences between
the white corpuscles on the one hand, and the lymph and
splenic corpuscles on the other, as regards size, colour, form,
and structure, and then proceeds to the discussion of the
" granule sphere " (Körnerkugeln). Here he shows that the
granule balls of Semmer mainly disappear when blood is
allowed to coagulate, " being simply the more coloured corpuscles of the *fugitive group*. They melt down into fibrin
without undergoing granulation, but when by the use of cold
this is prevented they undergo granulation and show themselves
as coloured granule spheres. When the corpuscles first break
up these granules are coloured, but they subsequently, *i.e.* in a
few hours, give up their colour to the liquor sanguinis and
appear white." Dr. Norris considered that the observations of
Semmer *when properly interpreted* lend no support whatever to
the theory which regards " the ordinary white corpuscle of the
blood as the precursor of the red disc." He also stated his
opinion that the hæmatoblasts of M. Hayem are in reality the
granules which result from the breaking up of the invisible
corpuscles when the blood is shed, which under these conditions
become visible.*

The third part is "On the Origin of the Mammalian Red
Corpuscle." Here he summarises his views on the whole
subject and declares that " there exist in mammalian blood
numerous corpuscles which are incapable of being seen by the
microscope, not because of their minuteness, but owing to the
fact that they have the same *refractive index* and *colour* as the
liquor sanguinis in which they are submerged. When brought
into view and carefully examined by suitable methods they prove
to be *colourless biconcave discs*, and between them and the red
biconcave discs biconcave corpuscles possessing every gradation
of tint can be detected." The morphological elements of the
lymphatic glands and spleen "*prove to be discs of the same size as*

Vide Section IV.—"An Examination of the Researches of M. Hayem
on the Development of Mammalian Blood."

the red corpuscles which are gradually becoming biconcave, the equivalent of these in the blood, the analogues of the white corpuscles of ovipara, being the invisible corpuscle and not the ordinary white corpuscle. Finally, the yellow and red granule balls of Semmer are produced by disintegrative changes in young, coloured, biconcave corpuscles resulting from the artificial conditions to which the blood has been subjected, and are entirely absent in perfectly fresh blood. In this paper the modes of tinting the liquor sanguinis and of staining the invisible corpuscle are also described. Solutions of sodium chloride charged with a colouring matter such as hæmoglobin or saffron are employed, and the following directions are given for the examination of the blood. " Place upon the end of the finger a small drop of the staining fluid, and with a needle prick the fingers through this drop, so that the blood may, when the finger is squeezed, flow directly into the liquid which has the double property of both preserving and staining." The invisible corpuscle is stained by a saturated solution of sodium chloride to which carmine and solution of ammonia have been added.

Dr. Norris's second paper entitled " Further Researches on the Third Corpuscular Element of Mammalian Blood," was read to the Society on the 10th of June, 1880. It is mainly a criticism of objections to his observations, though new. proofs of their accuracy are also offered, as regards both the nature of the third corpuscular element and its relations to the coagulation of the blood.

He begins by recalling attention to the fact that there exist in the blood of mammalia a very large number of colourless discs which had escaped recognition. " A colourless disc is now found in the blood corresponding numerically with, and which I shall be able to show has the properties of, a lymph corpuscle, and not those of a decolourised red disc, and between this and the fully-matured red corpuscle others, having every intermediate shade of yellow tint, can be observed," Vol. ii., page 197. He states his view that the biconcave corpuscles do not become visible at all in the blood till they have acquired just that slight amount of colour which is necessary to enable them to be distinguished from the liquor sanguinis, and that below such barely visible corpuscles there are, therefore, numbers wholly invisible, and the palest of the visible order may, under the influence of certain causes, lose their colour and join the invisible group.

Dr. Norris considers that he is able to divide the biconcave discs into three sets. 1. The primary group, including the whole of the colourless discs, and such of the coloured as are less tinted than those of the secondary group. 2. The secondary group of lustrous flickering corpuscles, barely capable of maintaining themselves in the absence of liquor sanguinis, and

having nearly as much colour as those of the tertiary group.
3. The tertiary group, including all corpuscles which do not
become lustrous, and are able to maintain a distinct outline in
the absence of the liquor sanguinis. " These primary groups
are often to be seen undergoing conversion into fibrin. The
corpuscles of these groups are *de facto* fibrin, and the delicate
fibres and layers which appear on glass slides are due, first, to
the extension of these granulations into fibres, or to annulation
of the entire corpuscle; or, secondly, to the spreading and laying
down of these corpuscles into films," page 214. When
blood is completely defibrinated these corpuscles and their
granules entirely disappear. In reference to the words " these
granulations" the author had previously drawn attention
to one of the photographs which shows the commencing
granulation of masses formed by the coalescence of the
corpuscles, themselves afterwards separating into distinct
granules. The colourless discs, however, are also converted
into fibrin without passing through the stage of granulation.

He shows the various characters of the corpuscles in
respect of colour, liquidity, granulation, relation to stains,
and specific gravity. As regards the last character it is stated
that, " like the white corpuscle, they are lighter than the red,
and have a tendency constantly to rise to the surface of the
blood, consequently the largest numbers are always seen to
attach themselves to the upper glass in preference to the lower,
and especially if time is allowed them to rise. This, no
doubt, has something to do with the *buffy coat.*" As regards
granulation it is stated that while the red corpuscles rarely
undergo granulation, these can scarcely be prevented doing so,
and that they may be readily stained by a weak solution of
aniline blue in a three-quarter per cent. solution of common
salt. Dr. Norris again describes his mode of examining the
blood in the following words :—" If we place upon the tip of
the finger a minute drop of saturated solution of salt, and prick
through it so that the blood may flow directly into the saline
solution, the refractive power of the liquor sanguinis is modified, and it is found that if we run this mixture of salt and
blood between glasses prepared according to the packing method
before described, we can then see the *outlines* of the colourless
discs, and the clear spaces, which have hitherto been supposed
to consist of liquor sanguinis only are observed to teem with
these discs."

The paper ends with a summary of reasons why the
invisible colourless discs cannot be regarded as decolourised red
discs ; among the rest, because they have neither the physical
nor chemical constitution of decolourised red discs, but of
lymph or gland corpuscles ; moreover, the disintegrative

changes which take place in these corpuscles give rise to the formation of fibrin in the blood; and the fibrin which is formed in the lymph has its origin in similar changes in the gland corpuscles.

In conclusion, it is desirable to state that a short abstract of Dr. Norris's first paper was given in the 9th vol., 1st part, of the " Jahresberichte über die Fortschritte der Anatomie und Physiologie." The Reports of this Society are given on page 26 as the source of this paper; but the immediate reference is to the " Centralblatt f. Med. Wiss " Nr. 22, S 402. The abstract itself, on page 32, gives the more salient points of the research, and especially alludes to the transition-stages (Uebergänge) in all stages of colour between the biconcave colourless discs and the perfect red corpuscles.

DIE BLUTPLÄTTCHEN DER SÄUGETIERE UND DIE „ INVISIBLE CORPUSCLES " VON NORRIS.

Von Prof. G. Bizzozero in Turin.

" Nachdem die Zeitschrift „ The Lancet " (No. 3, 21 Januar) meine der Turiner medicischen Akademie vorgelegte und in diesem Blatt (No. 2, 1882) veröffentlichte Mitteilung über die Blutkörperchen wiedergegeben, richteten zwei englische Aerzte, die DD. Neale und Norris, an den Herausgeber der „ Lancet " zwei Briefe (No. 4, 28 Januar), worin sie die Priorität der Entdeckung für sich in Anspruch nehmen. Dr. Neale erzählt, dass er bereits im J. 1854 im Blute kröpfiger Subjecte Körperchen beobachtete, die ihm identisch scheinen mit den in meiner Arbeit beschriebenen Gebilden. Nun kenne ich zwar seine diesbezügliche Original-Mitteilung nicht, doch was er jetzt hierüber schreibt, beweist zur Genüge, dass dieselbe nichts mit dem Gegenstande meiner Untersuchungen gemein hatte. Ich habe nie behauptet, ich sei der Erste gewesen, der im frisch entzogenen Blute, aufser den roten und weifsen, andere Körperchen gesehen hätte. Solches haben in der Tat schon vor mir andere Forscher beobachtet und habe ich auch nicht unterlassen, die Namen derselben gehörigen Orts anzuführen. Was aber, meines Wissens, vor mir noch Niemand (auch nicht Dr. Neale) getan, das ist, den Nachweis zu liefern, dass wirklich im circulirenden Blute der lebenden Säugetiere aufser den roten und weifsen Blutkörperchen noch typische Formelemente einer dritten Art enthalten seien, als welche sich

eben meine Blutplättchen herausstellen, deren Eigenschaften
ich studirt und auf deren Bedeutung für die Thrombosis und
enge Beziehungen zu der Blutgerinnung überhaupt ich in der
erwähnten Mitteilung hingewiesen habe.

"Derselbe Einwand gilt gegen Dr. Norris, der im J. 1879
gewisse Körperchen beschrieben hat, welche normal im Blute
enthalten, für gewöhnlich aber unsichtbar sein sollen*). Auch
seine Angaben können nicht Stich halten gegen den Vorwurf,
den man solchen Untersuchungen zu machen pflegt: dass
nämlich derartige Gebilde nur Zersetzungsproducte des Blutes
seien, die entweder der Entziehung als solcher oder der ange-
wandten Präparirmethode ihre Entstehung verdanken. — Mit
dieser Bemerkung könnte ich füglich meine Discussion mit
Dr. Norris abschliefsen, wenn er nicht an einer Stelle seines
Briefes auch noch darauf Prioritätsansprüche erhöbe, die Beziehungen zwischen den angeblich von ihm entdeckten Körperchen
und der Gerinnung des Faserstoffes herausgefunden zu haben,
weshalb ihm meine Untersuchungen nur eine ,,most important
and valuable confirmation" seiner Ansichten abzugeben
scheinen. Demgegenüber sei es mir gestattet, meine Meinung
über den inneren Wert seiner Entdeckung auszusprechen.

" Vor Allem fragt es sich, ob meine Blutplättchen identisch
seien mit den Norris'schen Körperchen ? — Nach N.'s
Beschreibung sind diese letzteren von derselben Größe, wie die
roten Blutkörperchen und verlieren nur in dem Maaße ihre
sonstige Unsichtbarkeit, als sie sich allmählich mit Hämoglobin
anfüllen, um sich zu roten Blutkörperchen zu verwandeln.
Dagegen besitzen meine Blutplättchen gewöhnlich einen 2—3
Mal kleineren Durchmesser, als die roten Blutkörperchen und
sind nie durch Hämoglobin gefärbt. Sie sind sichtbar, weil sie
etwas körnig sind und ein etwas anderes Brechungsvermögen
besitzen als das Blutplasma, während die N.'schen Körperchen
in gleicher Weise, wie letzteres, das Licht brechen und daher
nur dann sichtbar werden, wenn sie sich leicht gefärbt haben.
Allerdings könnte N. (wie er dies bereits in Betreff der
Hayem'schen ,, Hematoblastes" getan) die vermeintliche
Identität von beiderlei Elementen durch die Annahme zu ver-
fechten suchen, es seien die Blutplättchen ein Alterationsproduct
seiner unsichtbaren Körperchen. Doch könnte ich diesen
Einwand sehr leicht durch den Hinweis auf die Tatsache wider-
legen, dass die Blutplättchen mit ihren typischen Kennzeichen
im lebenden und kreisenden Blute, also unter ihren normalsten
Bedingungen, sichtbar sind.

" Diese Gegenwart der Blutplättchen im kreisenden Blute
stellt ihr wirkliches Vorhandensein beim lebenden Tiere aufser

---

* Cbl. 1880, S. 402.

aller Discussion. Können wir dasselbe von den N.'schen Körperchen aussagen ?

"Ich las aufmerksam die Beschreibung der Methoden, welche N. bei der Darstellung und Untersuchung seiner Körperchen in Anwendung gezogen hatte und ich wiederholte mehrere seiner Beobachtungen, mich streng an seine Anweisungen haltend. So habe ich auf's Bestimmteste die von ihm gemeinten Gebilde sehen können und bewahre die betreffenden Präparate, die ich Allen, die sich dafür interessiren, zu zeigen gern bereit bin. Die Ergebnisse meiner Untersuchungen lauten dahin, dass die von N. beschriebenen Körperchen nichts anderes sind als rote Blutkörperchen, die besonders durch die Manipulationen beim Präpariren ihr Hämoglobin verloren haben und daher unter Beibehaltung ihrer Form und Gröfse durchsichtig und farblos geworden sind, wie sie N. eben schildert. Unter solchen Umständen ist es begreiflich, dass bei einiger Uebung in der Darstellungsmethode man nach Belieben aus ein und demselben Blute Präparate gewinnen kann, welche arm, reich oder überreich sind an den fraglichen Körperchen.

"Nach Alle dem darf ich wohl auf's Entschiedenste erklären, dass die NORRIS'schen ,, Invisible corpuscles" gar nichts mit meinen Blutplättchen zu schaffen haben ; und schien mir diese Erklärung, zur Vermeidung allen Missverständnisses, um so dringender geboten, als mir daran gelegen ist, meine Untersuchungen über das Blut recht bald und ernstlich von anderen Forschern geprüft und bestätigt zu sehen."

NORRIS AND BIZZOZERO.

From the " Lancet," March 18th, 1882.

We have received from the secretaries of the Birmingham Philosophical Society an extract from the minutes of a meeting on February 9th, which contains a brief historical sketch of Dr. Norris's remarkable papers on the blood, which are contained in the Transactions of the Society in the last four years. Attention was called to the subject by Dr. Heslop, the President, on account of the discovery by Bizzozero of a new bloodcorpuscle, of which we gave an account in our number for January 21st. In our article we pointed out the resemblance of these corpuscles to those which have been repeatedly described by Norris. There appeared to be a *primâ facie* difference between the two in size, and in the fact that the special

characteristic insisted on by Norris is that his corpuscles have
a refracting power so nearly that of the plasma of the blood,
that they are invisible until brought into view by artificial means
whereas those of Bizzozero were actually seen by him in the
circulating blood. The closeness of the resemblance between the
two, is, however, clearly shown by the facts mentioned by Dr. Norris
in the letter from him which we published on January 28th, and
by the facts mentioned in the Birmingham Society's minutes.
It is urged by Dr. Norris that the " new corpuscle " is only the
"invisible corpuscle" in the further stage of development
which he has traced, in which it undergoes such changes as to
become visible without artificial aid.* Dr. Norris further pointed
out that the statements of Bizzozero as to the probable share
taken by these elements in the formation of fibrin, and the
method by which they may be rendered more distinct by
colouring agents, are practically identical with his own.

To this Professor Bizzozero replies in the current number
of the *Centralblatt für die Medicinischen Wissenschaften.* He first
alludes to the letter in which Dr. Neale refers to an early
observation by himself of similar elements in the blood of
goitrous patients. Bizzozero denies any intention of claiming
to be the first to observe other corpuscles in the blood besides
the red and the white, since such bodies have been noticed by
many other observers, but what he believes that no one before
himself has done is, to see in the circulating blood of living
mammalia, besides the red and white cells, a third form of
typical elements such as the "Blutplättchen," to have studied
their peculiarities, and to have established their relation to the
process of coagulation.

Bizzozero then asks the question: Are his "Blutplättchen"
identical with the corpuscles of Norris ? The latter are described
as similar in size to the red discs, and to lose their invisibility
when they gradually become charged with hæmoglobin in the
process of transformation into red discs. On the other hand,
the corpuscles of Bizzozero are usually one-half or one-third the
size of the red corpuscles, and never become coloured with
hæmoglobin. They are visible, because they are slightly
granular, and possess a slightly different refracting power from
the blood-plasma. To the suggestion of Norris that the "Blutplättchen" are a product of the alteration of his corpuscles,
Bizzozero responds that this cannot be, since the "Blutplättchen"
are visible in typical form in the living and circulating blood
under its most normal conditions. He states that he has
repeated Norris's experiments, and has seen the corpuscles he
describes most distinctly, but believes that they are simply red

* This is a slight misconception of my view. *Vide* communication
in the "Lancet," April 8th, 1882, page 250.

corpuscles, which have become partially or wholly decolourised by the method of manipulation employed. This allegation we may fairly leave in the hands of Professor Norris ; but we may remark that few who witnessed his brilliant demonstrations, given a few years ago at the College of Physicians and elsewhere, will be disposed readily to accept this explanation of the origin of these bodies, whatever their nature may be. Nor does the objection that the " Blutplättchen " cannot be derived from the invisible corpuscles, because they are seen in the circulating and normal blood, appear altogether satisfactory. To the close resemblance between the opinions expressed by Norris and himself regarding the relation of their respective corpuscles to the process of coagulation, and regarding the means by which these elements may best be rendered visible, Bizzozero does not allude.

ON THE CLAIM OF PROFESSOR BIZZOZERO TO THE DISCOVERY OF THE FIBRIN-FORMING CORPUSCLE OF THE BLOOD.

By RICHARD NORRIS, M.D. St. And., F.R.S.E.,
PROFESSOR OF PHYSIOLOGY, QUEEN'S COLLEGE, BIRMINGHAM.

From the "Lancet," April 8th, 1882.

In a leader in "The Lancet" of March 18th, reference is made to a communication by Professor Bizzozero, which appeared in the "Centralblatt" of March 11th, in reply to a letter of mine, published in "The Lancet" of January 28th. It seems to me that, quite apart from all personal considerations, the question is one of such general and paramount interest in physiology and pathology, as to warrant me in returning to the subject, with a view to the more complete elucidation of its existing position. It is a well-known fact, of which I possess the fullest documentary evidence, that so early as February, 1878, I sought to secure to British physiology any credit which might accrue from the discovery of the fact that the existing chemical views of coagulation were destined to give place to morphological ones ; in brief, that there existed in the blood a corpuscle, the degenerative changes of which were competent to account for and explain all the phenomena of fibrin formation. I have traced these corpuscles, step by step, throughout the whole of their changes and variations, and have photographed each successive divarication, till they present themselves in the well-recognised forms of fibrin—e.g., films, networks, embolic masses, &c. Now, Professor Bizzozero claims for the corpuscles which he has seen the same properties which long ago I ascertained to belong to my corpuscles, nevertheless, he denies to me the discovery of any *new body*, and contents himself by affirming that I have mistaken *decolourised discs* for a new element of the blood. Where is the logic of this? Will Professor Bizzozero maintain that, having first deluded myself, I have next entered into all kinds of speculations relative to the fibrin-forming power of this body, and that these speculations turn out so accurate as to render it simply necessary to discover the actual body to which they are related, and, by a simple exchange, at once to complete an important research, the final issues of which had been already worked out; for there is no new fact in the original communication of Professor Bizzozero, save that he has seen in the vessels of chloralised rodents a corpuscle with which he was not previously familiar? It seems to me that the only other alternatives open are, either to admit that decolourised discs are the fibrin factors of the blood, or to affirm that he has also discovered in the blood another and an independent corpuscle, which is capable of undergoing the same series of changes as mine. Is not the suspicion, which has already occured to Professor Bizzozero, that he has come across one of the many transition phases of the corpuscles which I have described much more probable? I entirely concur with him when he says: " Above all, we must know if my blood-discs are identical with the corpuscles of Norris." Proceeding to contrast these corpuscles, he says: " The latter are, according to the description of Norris, of the same magnitude as the red discs, and lose their former invisibility only in the measure that they become charged with hæmoglobin; on the other hand, my blood-discs have generally a diameter two or three times smaller than that of the red corpuscles, and are never coloured by hæmoglobin; they are visible because they are granular, and because they possess a power of refraction differing from that of the plasma of the blood, whilst those of Norris refract the light in the same way as the plasma, and become therefore visible only when they are slightly coloured." He proceeds, " Truly, Norris might try to defend the supposed identity of both elements on the supposition that the discs might be a transition product of his invisible corpuscles, yet I might easily disprove this supposition by showing that the discs are visible, with their typical characteristics in their most normal condition in the living circulating blood itself." The description here given by Professor Bizzozero of the bodies he has observed is much more

explicit than in his original communication, in which he said
that the reason they had hitherto escaped notice was due to
their being colourless and translucent, but he now adds that
they are visible because they are granular, and because they
possess a power of refraction differing from that of the plasma
of the blood. We have now, to guide us in their identification,
five things: size, form, visibility, granular character, and
absence of colour.

At this point it is desirable for me to state again my case,
in a manner which does not seem to have presented itself to the
mind of Professor Bizzozero. The question at issue is not
whether the discs described by him are identical with my invisible corpuscles, but rather whether they belong to that series
of discs which, on account of their tendency to degenerate into
fibrin, I have designated the "fugitive group," some of the
corpuscles of which are visible, others barely visible, and others
still wholly invisible. At both ends of this group there are
visible discs, for it includes on the one hand the advanced
or nuclear lymph-discs, which are antecedent to the invisible
discs, and on the other those discs which have again become
visible by the acquisition of colour, but which are not yet
sufficiently stable to be included in the permanent group of
red discs. These limitations I have repeatedly stated in my
papers. According to my view, therefore, all the younger discs
of the blood, some of which are colourless and yet visible (like
the white corpuscles, but more delicate), others of which are
colourless and invisible, and others partly coloured, and therefore visible, undergo conversion into fibrin when the blood is
shed. Professor Hayem is well acquainted with the fact that
bodies containing colour contribute to the formation of fibrin,
but he has not yet realised that these bodies are fragmentary
parts and variously modified forms of the partially coloured
blood-discs, hence he has described them as the independent
bodies which he designates hæmatoblasts. I have taken great
pains to prove that the so-called lymph-globule is, in reality,
a disc-shaped body—a little smaller, but, at the same time,
slightly thicker than the blood-disc, and that it consists of two
varieties, to which I have given the names " primary " and
" advanced," the former being a true cell of about the same
diameter as the red blood-disc, or slightly less, but having a
greater thickness ; the capsule of this cell is intimately applied
to the nucleus, but can easily be demonstrated by osmosis.
The advanced lymph-disc is the delicate, naked nucleus of this
body ; both these bodies exist in certain numbers in the blood, in
a *visible* state, and the latter is the body which is resolved into
fibrin in the lymph, as its derivative, the colourless invisible disc,
is in the blood. These lymph-discs are poured into the blood

in large numbers, at the subclavian and splenic veins, and for the most part, being colourless and already smooth, enter at once upon the stage of invisibility; but those among them which are not so perfectly elaborated remain for a time visible, and are at this stage colourless, finely granular discs, slightly smaller than the red discs. This appears to me to be the body to which Bizzozero refers, and as it forms fibrin while still in the lymph, so, *à fortiori*, when present as one of the elements of the "fugitive group" in the blood, it is a fibrin-forming corpuscle; but the difference between the fibrin formed by this body and the invisible and intermediate discs is precisely the difference between the coagulation of the lymph and of the blood—the one yielding, soft and granular, the other more homogenous, viscous, and tenacious fibrin. These bodies, for many years familiar to me, must have been seen by most observers of the blood, but regarded as delicate minute forms of the white blood-corpuscle. It would indeed be a matter of surprise if they could not be seen in the circulating blood when carefully looked for, and therefore the fact that " they are visible with their typical characteristics in their most normal condition,· in the living circulating blood itself," is not the slightest disproof of the view that they are transition forms. They are, in fact, a slightly less developed stage of the invisible colourless disc which eventuates in the red disc, and their gradual transformation into this body may be traced both in the lymph and in the blood. I have already incidentally referred to the allegation of Professor Bizzozero, that the invisible corpuscles of the blood as displayed by my methods are decolourised red discs. Such a view naturally occurs as a possible contingency to the mind of every one, and it was the first point to the clearing up of which I seriously applied myself. Professor Bizzozero will pardon me for saying that he is at present only on the threshold of this research, and occupies to-day the position in which I found myself in the latter part of the year 1877. He seems to be unacquainted with my paper, published in the Proceedings of the Birmingham Philosophical Society in June, 1880. in which I deal exhaustively with this aspect of the question, and bring together numerous overwhelming proofs that this facile explanation is altogether untenable.

The singular notion that blood-discs can discharge the whole of their hæmoglobin and become suddenly (within ten seconds) colourless, not only to the eye, but to infinitely more delicate tests, of a photo-chemical nature, and that, in a serum, which must (if the corpuscles lose it) necessarily contain hæmoglobin, or, as is the case, in a serum which has been purposely saturated with hæmoglobin (derived from an independent source) at the moment it leaves the vessels, is one entirely of modern growth—in fact, has been invented to meet my discovery. If to draw a drop of blood from the end of the finger, and to place it on a slide, either at the temperature of the room, or at that of blood or ice, and to lay down upon it, in the gentlest manner, a flexible mica-cover, is a manipulation calculated to bleach to whiteness some corpuscles, while leaving others and contiguous ones, under precisely the same conditions, absolutely unmodified, then certainly there is an end to all such processes as corpuscle enumeration with its diluents, measuring tubes, cells, etc. That able investigator of the blood, Professor Gulliver, who, I believe, was the first to inquire into the question of the decolourisation of red discs, favoured me in 1879 with a letter referring to my first paper, in which he says, " It gives me pleasure to see that you support my original view as to the essential difference between lymph-corpuscles and the pale globules of the blood (in reality the very point now again raised from the blood side by Professor Bizzozero); your new corpuscle bears a great resemblance to the colourless basis of the red corpuscles. I described and figured the 'membranous bases,' and taught how to obtain them, upwards of a quarter of a century since, but this does not affect your discovery of the free and hardly seen corpuscles in the living blood ; indeed, if confirmed, your observation will have important significance." Clearly Professor Gulliver does not entertain the idea that corpuscles can be reduced to the stroma-form by the mere arrangements necessary to place blood under the microscope for examination. I might retort upon Professor Bizzozero that to examine blood-corpuscles in a solution of chloral or chloroform (for this is what happens with chloralised animals) is not nearly so normal a condition as to examine them within a few seconds of being shed, from the end of one's own finger, in their normal undrugged plasma. Such objections raised to the simplest methods, while conclusions are accepted without hesitation, when it is known that the blood contains foreign substances capable of maintaining an animal in a profoundly narcotised condition for hours, seem to me, to say the least, " straining at a gnat and swallowing a camel." Further, I have ascertained that it is not the three-quarter per cent. saline solution alone which prevents decolourisation of the corpuscles, but the colloid matter which the blood contains, and I find that by increasing this element, by adding a little dry soluble albumen to the blood, I can prevent exosmose of colouring matter from the corpuscles to such an extent that no change in them, or in the liquor sanguinis, is obvious after a period of eighteen hours. The invisible corpuscles are present as usual. Where, I ask, is the colour from these corpuscles gone to on the hypothesis that they are decolourised discs ? It

is not present in the liquor sanguinis, for this photographs as
clear and colourless as the adjacent air-bubbles. Numerous
other reasons are given in my paper, each one of which, even
when taken by itself, is decisive on this question. But to return
to the main issue, what is to be done with the cardinal and indisputable fact that these identical corpuscles can be traced step
by step into the well-known fibrin forms? In conclusion, I
would respectfully ask Professor Bizzozero, and physiologists
generally, to suspend judgment for a short period till my work
on the " Physiology and Pathology of the Blood," now on the
verge of publication, is issued, as it contains the evidence which
I have accumulated upon these questions during the past few
years.

INDEX.

Myelin, annulating power of, 32
— biconcave form probably due to, 31-32
— comparison of blood rings and rings of, 33-36c
— presence of in lymph corpuscles, 32
—— blood discs
Mussy, de Noel Geneau, observations on the researches of the author, 60

NELSON'S opaque gelatine, observations with, 31
Neumann on the development of the blood, 132
—— bone-marrow, xxxviii
Newton's rings, barrier of, how to form, 3
— degree of approximation of at filtering part, 3
Normal form of invisible discs, 8
Normal temperature, effects of on so-called hæmatoblasts, 80
— best preserves young or fugitive discs, 96
Nucleated appearance of granule spheres, 23
Nuclei of white corpuscle, probable destiny of, 124, 191

OBJECTIONS to premature designations, xliii
Observation of blood in single layers, results of, 2
Origin of colourless discs, 13, 40
Origin of red discs, 14, 15
—— white corpuscles, 120, 121, 124
Organs classified as "lymphoid," 193
Osler, white masses or sheets seen in blood by, 86
Osmic acid, action of dry vapour of, 8
— colourless disc partially preserved by
— decrease of so-called hæmatoblasts by, 91-92
— effects of cold combined with vapour of, 8
— form of red disc preserved by, 8
— mode of using ice-cold temperature in conjunction with, 9

Osmic acid, prevents fibrin-formation, 92
— red discs rendered insoluble in water by, 70
— red discs contracted or condensed by
— shallow pan for using vapour of, 8
Osmosis, use of to display cell walls, 117
Ovipara, function of lymphoid organs of, 194
—— coagulation in the blood of, xxxv
—— permanent and fugitive corpuscles of, xxxv
Oviparous blood, analogue of invisible disc of mammal in, xliii., 183
— analogue of nuclei of white corpuscles in, 186
— apparent free nuclei in, 183
— derivation of bodies which form red corpuscles in, 186
— difference between M. Pouchet's view and the Author's of development in, 188
— function of nucleolated nucleus of white corpuscles of, 187
— hyaline-bordered nucleated corpuscles, how to demonstrate in, 186
— investigations of M. Pouchet on development of, 187
— mode of growth of red corpuscles in, 186
— nucleated elliptical cell of, 183
— origin of colourless elliptical cells of, 183
— origin of white corpuscles of, 188
— pseudo-gemmation in embryo, xl
— range of corpuscular development in, 188
— red nucleated corpuscle, minor mode of development of, in, 185
— so-called hæmatoblasts of Hayem in, 186
— use of Table I. and Diagram II., Plate xx. in explaining development of, 190, 191, 192
Oviparous bone-marrow, colourless nucleated ellipsoids of, 183
— comparison of mammal bone-marrow with, 184
— difficulty of demonstrating colourless ellipsoids of, 184
— effects of stains on colourless ellipsoids of, 184
— major process of blood production carried on in, 185

IN COURSE OF PREPARATION,

A further research on the Coagulation of the Blood, in which the microscopical and chemical aspects of the question are fully discussed, and the existing physical and chemical facts and phenomena shown to be in harmony with morphological views of fibrin formation.

THE HERALD PRESS BIRMINGHAM

FORWARD